The 15 Most Common Obstacles to World-Class Reliability:

A Roadmap for Managers

Don Nyman

Industrial Press

Library of Congress Cataloging-in-Publication Data

Nyman, Don.
 15 most common obstacles to world-class reliability : a roadmap for managers / Don Nyman.
     p. cm.
  ISBN 978-0-8311-3381-8
 1. Quality control. 2. Production management. I. Title. II. Title: Fifteen most common obstacles to world-class reliability.
  TS156.N96 2009
  658.4'013--dc22

                                    2008054165

Industrial Press, Inc.

989 Avenue of the Americas
New York, NY 10018

Sponsoring Editor: John Carleo
Copyeditor: Robert Weinstein
Interior Text and Cover Design: Janet Romano

10   9   8   7   6   5   4   3   2   1

Table of Contents

Preface

Regardless of industry, organizations are responsible for optimized utilization of asset capacity reflective of invested capital. The goal is for capital assets to achieve the reliable output capacity for which they were designed, and on which they were justified, allowing the business entity not only to survive … but to **thrive**! This quest is crucial to "Return on Invested Capital" — that financial indicator on which Wall Street places so much emphasis.

Historically, whenever sales demand increased dramatically, there was a strong tendency to construct additional capacity (more capital investment — more bricks and mortar) even if design capacity of existing plants was not being realized due to reliability shortfalls. Although elusive, Industry's current emphasis upon **Reliability** is long overdue. This book is recommended to all management teams frustrated in their quest for realization of the design capacity on which their capital investments were based.

Many organizations believe they are pursuing World-Class Reliability, yet continue to operate in a reactive environment, perpetually reacting to emergencies and doing little to eliminate root causes.

**The Challenge is
World-Class Reliability
Supportive of
Business Objectives!**

The Purpose of This Book

The purpose of this book is to clarify for management the obstacles commonly encountered in its quest of Reliability. Throughout my fifty-year career supporting hundreds of organizations, the blame for reliability shortfalls have most often been placed upon the Maintenance Department; which has been commonly regarded as a "necessary evil and drain on profitability." In reality, although most reliability initiatives are maintenance dependent, there is plenty of blame to be shared across a broad spectrum of organizational functions, top-to-bottom, inclusive of most functions.

Fifteen obstacles to the achievement of World-Class Reliability are consistently identified whenever comprehensive, maintenance/reliability assessments are conducted. These obstacles relate largely to cultural and environmental habits that upper-management must change before more specific reliability efforts by middle managers (line and staff) can be implemented and inculcated. Management must establish the atmosphere that provides life-sustaining oxygen to World-Class Reliability efforts. Thereby, the focus and uniqueness of this book is emphasis on the managerial leadership required to transform reactive operating environments to proactive environments through culture change, universal commitment, and perseverance. Such transformation is essential if an organization is to **optimize asset reliability**. The struggle for integrity between culture and the reliability mission must never wane.

Some management readers may minimize the significance of the identified obstacles. However, during many years presenting Maintenance/Reliability Excellence seminars, I've found that a common reaction of maintenance functionaries is:

"You are preaching to the choir. **We** recognize these as real barriers to realization of reliability goals. **Does upper management? We** cannot effectively address them at our level!"

Those organizations that have already achieved World-Class Reliability have obviously hurtled the obstacles. However, before you judge yourself to be in that extremely small august group, it may be well to read on.

The book does not provide comprehensive coverage of the many approaches and tools deployed in pursuit of reliability such as Preventive/Predictive Maintenance (PM/PdM), Total Productive Maintenance (TPM), Reliability Centered Maintenance (RCM), and Planning and Scheduling (P&S). These are the vehicles used to identify and pursue opportunities by which to improve reliability. However, highly regarded books already exist on these subjects (see Supplemental Reading). Here they are postured within broader messages, but are not covered at the practitioner level. Although upper management need not understand the application of the tools, they must comprehend that full potential will not be realized or sustained until the barriers addressed within this book are hurtled.

For three decades, clients have paid significant professional fees to access the insights presented herein. As retirement approaches, I do not want the insights, experience, and knowledge to be lost. For the nominal cost of this book, the insights are offered to all. The potential gain is great for those willing to carefully read and absorb the materials offered and, accordingly, guide renewed reliability actions.

Donald H. Nyman
Hilton Head Island, SC
2009

Introduction

All organizations are responsible for effectively spending each dollar invested in maintenance labor and material in order to achieve equipment reliability supportive of the asset capacity required to meet business objectives. Product quality, on-time delivery, regulatory compliance, optimal profitability, and return-on-asset value are all dependent upon reliable output to the limits of designed capacity. None of these objectives is achievable in the state of reactive confusion that inevitably exists when operations are focused only on short-term survival (Quarterly Financial Results).

The basic need is for World-Class Operations built upon reliability supportive of business needs and objectives. World-Class Operations make an organization equal to or better than the best in the world of international competition. The pathway to this vision is Maintenance/Reliability Excellence, predicated upon pro-action, not reaction. The Reactive Cycle must be broken before Reliability can be achieved.

Although many organizations claim to be in pursuit of World-Class Reliability, they continue to operate within reactive cultural environments wherein Maintenance resources are organized primarily for reactive response. All too often, urgent responses to repetitive failures consume the bulk of available

Maintenance Manpower to the detriment of proactive efforts designed to identify and eliminate root causes of failures. These same organizations continually wonder why they live in a state of perpetual reaction to breakdowns and other emergencies. Allowing the same failures to continue repeatedly without determining and addressing their root causes is egregious mismanagement. As Benjamin Franklin and Albert Einstein both stated:

> "Continually doing the same things over and over, expecting different results … is insanity!"

The common obstacles that prevent many organizations from achieving reliability lie in lack of maintenance understanding, which stems from a lack of appreciation for the maintenance function.

The fifteen common managerial obstacles to achievement of World-Class Reliability explored within this book are organized into eight chapters:

Chapter One — Creating a Culture for Reliability (page 1)

This chapter speaks to Upper Management and addresses the proactive environment that must be established for Reliability to be achievable.

- Obstacle 1 Lack of Understanding, Beliefs, and Commitment Shared by the Entire Organization (page 1)
- Obstacle 2 Lack of Integrated Missions, Plans, and Incentives (page 5)

Chapter Two — Structuring the Maintenance Organization for Reliability (page 29)

If reliability is to be achieved and sustained, Maintenance must be organized for Reliability. This chapter is directed primarily at senior Maintenance and Operational Managers.

Chapter Three — Clarifying Required Asset Capacity and Associated Maintenance Resources (page 37)

Available asset capacity is theoretical unless provision is made for essential maintenance. The prime mover in this regard is Reliability/Maintenance Engineering.

Chapter Four — Balancing Maintenance Resources with Workload (page 45)

• Obstacle 10 Imbalance of Maintenance Resources with Maintenance Workload (page 46)

Chapter Five — Skills Required to Maintain Advancing Technology (page 55)

• Obstacle 11 Skills of Many Maintenance Personnel Fall Short of Levels Required to Maintain Advancing Equipment Technology (page 56)

Chapter Six — Materials Support (page 65)

• Obstacle 12 Procurement is Focused Too Much on Initial Cost versus Life Cycle Cost (page 66)

Chapter Seven — Preparation of Maintenance Work (Page 83)

A planned job requires only half as much downtime as an unplanned job; and each dollar invested in preparation saves three-to-five dollars during work execution. Workers do not plan for their own efficiency. The prime movers here are Planner/Schedulers and Supervisors.

• Obstacle 13 Insufficient Preparation for the Execution of Maintenance Work (page 83)

ORIGIN OF IDENTIFIED OBSTACLES AND INSIGHTS OFFERED

Most of my fifty-year career has been devoted to the improvement of maintenance effectiveness for hundreds of organizations across a variety of industries. For many of those organizations, comprehensive assessments were conducted of their maintenance effectiveness. Initial assessments led to master plans for improvement. Annual follow-up assessments identified accomplishments and remaining short falls — with reasons for the latter and action steps by which to re-ignite positive momentum. The extreme commonality of conditions observed lead to the 15 Obstacles postulated within this book.

Some clients achieved World-Class Maintenance and Reliability. Others are well on the way. Many made initial progress but lost focused commitment.

"Stand up to your obstacles and do something about them.
You will find that they haven't
half the strength you think they have."

Norman Vincent Peale

Acknowledgements

As always, I am most appreciative of the endless encouragement and support of my wife Barbara and daughters Laurie and Sharie.

I am indebted to my long time friend and collaborator, Pete Little; and to my two sons, Jeff and Ture. Pete edited the original manuscript, made helpful comments and contributed to the discussion of skills training essential to achieving and sustaining reliability of high technology processes and equipment. Jeff and Ture contributed to the discussion of materials management. As suppliers to industry, their insights were valuable.

I am grateful to the support group at Industrial Press: John Carleo, Suzanne Remore, Janet Romano, Robert Weinstein, and Daniel Daley. Without their efforts the book may not have made it to print.

Kevin Lewton of METDEMAND suggested the concept of Management creating the atmosphere that provides life-sustaining oxygen to World-Class Reliability.

Contact Information:

Kevin Lewton	Pete Little
METDEMAND	MPACT Learning Center
Charleston, IL	Greensboro, NC
(888) 427-4330	(336) 379-1444
metdemand.com	Little.Pete@MPACTLearning.com

continued on next page

Ture Nyman
MRO Products
Mt. Pleasant, SC
(843) 568-4003
tnyman@comcast.net

Don Nyman
The Nyman Consulting Group
Hilton Head Island, SC
(843) 341-7677
nymandon@hargray.com

Jeff Nyman
Action Bolt & Supply
Rock Hill, SC
(803) 324-2658
actionbolt.com

Supplemental Reading

As stated in the Preface, this book does not provide comprehensive coverage of the many approaches and tools deployed by middle managers, engineers, and technicians in the pursuit of Reliability. Highly-regarded books exist on those approaches and tools. Some of the best are listed below for the benefit of those readers who wish to go deeper.

RELIABILITY CENTERED MAINTENANCE AND ROOT CAUSE FAILURE ANALYSIS

"The Little Black Book of Reliability Management"
Daley, Daniel T.
Industrial Press, New York; 2007

"RCM II" – Reliability-Centered Maintenance
Moubray, John.
Industrial Press, New York; 1997

Root Cause Analysis
Latino, Robert J. and Kenneth C. Latino
CRC Press, New York; 1999

WORK PREPARATION
Maintenance Planning, Scheduling & Coordination
Nyman, Don and Joel Levitt
Industrial Press, New York: 2001

CHAPTER ONE
CREATING A CULTURE FOR RELIABILITY

A significant majority of organizations (manufacturing and others) operate in a reactive mode. Their culture is inconsistent with their mission statements and with their lofty Reliability goal. For these organizations, cultural change must occur before significant reliability improvement can be achieved. This chapter addresses the six obstacles that upper management must address before bottom-up efforts of functional managers, supervisors, engineers, and planners can achieve and entrench Reliability.

OBSTACLE 1
Lack of Understanding, Beliefs, and Commitment Shared by the Entire Organization

Successful endeavors begin with beliefs shared by the entire organization. The origins of such beliefs are common understanding and appreciation. Therefore, organizations must be educated as to the principles, benefits, and contributions of "Reliability gained through Maintenance Excellence." Established principles (Appendix A) of maintenance management and reliability must be initially instilled through education and subsequently substantiated by bottom-line results.

In many organizations, maintenance is looked upon as a necessary evil, drain on profitability, non-contributor. This perception could not be further from reality, because most operational improvement and reliability initiatives are "Maintenance Dependent!"

The misperception stems from lack of: understanding and consequent appreciation for the maintenance function. Most managers and other organizational members have little-to-no maintenance experience. Often, even maintenance managers, supervisors, and technicians have never worked in anything but an environment of reactive repair. As a result, few members of the overall organization understand the proper functions of Maintenance. Nor do they appreciate the many contributions proactive maintenance can make to operational reliability, profitability, and business survival.

Reliability is not achieved by rapid response to daily problems, but by pro-active maintenance designed to eliminate or minimize those daily problems. Many people, even entire organizations, believe that "Repair is Maintenance." It is not! Repair is the consequence of failure to perform identified pro-active maintenance. The distinction is implied by the common functional name: Maintenance and Repair (M&R).

Maintenance is proactive. Notice that "M" comes first because it is the more important. Repair is reactive. To the detriment of reliability, the "R" commonly consumes a misdirected majority of skilled maintenance resources. Proactive maintenance minimizes reactive emergency response.

Once shared beliefs are established, the next step is to conduct an assessment to identify the gap between beliefs and current conditions. The assessment identifies and quantifies the "Maintenance Iceberg" (Appendices B and C) to establish the bottom-line financial justification and thereby gain management approval for pursuit of world-class reliability through maintenance excellence.

Identifying this gap also provides direction for a master plan of action steps with quantified goals supportive of operational and business plans, as discussed with Obstacle Two.

Process Steps to Address These Barriers

These and all process steps to be presented are reflected in the generic Master Plan presented in Appendix D. Readers may wish to follow down the Master Plan as each group of process steps is discussed.

1.Assess to Identify the Gap between Beliefs and Current Conditions.

Several consulting firms offer assessment support (see References: Appendix L). The availability of such support should be a selection requirement if an external seminar leader is necessary. If an internal resource exists, a suitable assessment process can be developed internally.

2.Initiate the Transformation Process to Proactive Reliability.

Begin with education. The initial challenge is to achieve a common understanding of and appreciation for the maintenance function and its relationship to reliability. This appreciation is best accomplished in a seminar format delivered to key members of the overall organization (see Appendix E for outline). A carefully selected external consultant is often retained to lead the local organization through the maintenance / reliability excellence process. However, if an internal resource with the necessary background exists, external support is not essential. Indeed an internal resource has the advantage of being both the up-front training facilitator and the on-going champion of maintenance/reliability excellence.

OBSTACLE 2
Lack of Integrated Missions, Plans and Incentives

All units of the organization share guilt for existing entrenchment of the reactive culture:

- Upper management espouses reliability but fails to budget and staff accordingly. Cost reduction is often emphasized to the detriment of reliability excellence.

- Production managers and supervisors are driven to meet daily production and shipping quotas to the detriment of reliability efforts that optimize longer term output potential (weekly, monthly, quarterly, annual).

- To the detriment of proactive work that is essential to sustained-reliability; maintenance managers, supervisors, and technicians are easily diverted to urgent demands because this is where they historically receive the most applause.

- Materials management is overly focused on the initial cost of new equipment and spare parts, rather than upon life-cycle cost that yields positive impact on reliability.

- Project engineering is focused on the installed cost of capital equipment without adequate provision for maintenance of said equipment (e.g., training, spare parts, periodic access for pro-active maintenance).

Stress and dissonance result due to functional incentives and resultant practices that are misaligned with aspirations of the overall enterprise. Covenantal partnerships should be emphasized and nourished.

Covenantal Partnership

Contractual partnerships tend to be reactive. As soon as one functional partner slips relative to one or more of its responsibilities, the other partners respond by neglecting their responsibilities to the common mission. Emphasis must be instead to covenants between parties rather than contracts. In covenants, functions forgive slippages by other parties and do everything in their power to live up to their own covenant promises to the overall partnership.

This characteristic is often the difference between Japanese team concepts and U.S. team concepts where frankly too many operational/maintenance relationships are contractual rather than covenantal. Cultural differences are again a factor in these differences.

Reliability of facilities, processes, and equipment is a responsibility shared by all members of the facility team, with complete interdependency among organizational units. Although each unit has its own purposes and objectives, no single unit can successfully function without the support of the other units. A spirit of covenantal partnership keeps all organizational units working together toward ultimate goals for the common good. When judgments are reached from this perspective, pro-activity expands while reactivity diminishes. Keeping overall objectives in the forefront allows resources to be scheduled based on broad objectives rather than being consumed by immediate crises. In contrast, whenever individual departments independently maximize functional interests, interests of the business entity are sub-optimized.

Although Operations is indeed an internal customer of Maintenance, as is Maintenance an internal customer of Purchasing and Stores, emphasis on partnership and covenant forestalls abusive use of the "customer is always right" syndrome. Herein, the term partner/customer is utilized.

Vision and Mission

Reliability through maintenance excellence is a continuous improvement process. It requires of management: sustained understanding, commitment, support, and involvement communicated via a well-conceived operational vision statement and related maintenance mission statement that together elicit a proactive environment and culture. It is essential that the vision be shared by the entire organization, guiding all units, in harmony, toward realization of ultimate business objectives. Without vision, day-to-day efforts tend to focus reactively on short-term targets of each individual organizational unit, which can be detrimental to ultimate, long-term goals of the business as a whole.

Development of vision and mission statements must be a cooperative, interactive process. It starts at the top with the overall vision. It works down and across functional units. Each function must convey its responsibilities to internal partners/customers and to external customers, together with its supportive responsibilities to other functions. The resultant package of vision and mission statement must be harmonious.

Mission statements are simply words. Yet, they are the stimulus that rouses the mind and spirit; thereby inciting pursuit of reliability. That said; remember that cultural change requires a lot more than words and speeches (and books).

Plans

Sales volume is dependent upon units produced; units produced are dependent upon reliability of installed capacity; capital plans influence installed capacity; and reliability of installed capacity is dependent upon compliance with maintenance plans. As a result, congruity of business, operating, maintenance, and capital plans is essential. Unless the four are mutually supportive, they are merely four independent and ineffective documents. Harmonious missions facilitate the development of integrated plans.

Goals and Incentives

Likewise, goals and incentives placed upon the various functional units must also be congruous. In today's corporate climate, extreme pressure is placed on each business function to capture and contribute all possible volume and cost reduction. In response, each function applies its own expertise to its own sphere of influence, as encouraged by functional incentives.[1] As a conse-

---

[1] Peter Drucker wrote; "Management often measures the wrong things." Here is a good example.

quence; industry is back on the traditional managerial treadmill. "Whenever individual functional units maximize their own interests, the interests of the business entity are sub-optimized." Keeping overall objectives in the forefront allows all resources (capital and human) to be invested optimally.

Spare parts are often purchased on price alone, which is seldom consistent with pursuit of reliability. Operations and Maintenance are dependent upon the reliability and longevity of the purchased item. World-class requires the search for the optimum in all situations. **Life-cycle cost analysis** is the established vehicle by which to achieve optimum in the example presented. Optimum does not maximize the selfish interests of any of the four individual functions (Engineering, Purchasing, Operations, and Maintenance), but it does optimize their mutual interests and those of the corporation.

Progress should be measured regularly. In turn, incentives awarded to the various organizational units should relate directly to measured progress toward eventual achievement of overall goals.

Clear Linkage of Organizational Responsibilities

Due to down-sizing throughout the last fifteen years or so, a number of critical functions have been decimated. Even when associated positions are reinstated, they are misused. They become a "jack-of–all-trades" rather than being focused on intended responsibilities. General responsibility statements for each functional partner are summarized below:

A. *Purchasing*

Purchasing sources, procures, and expedites delivery of all parts and materials.

B. *Stores*

The purpose of storeroom inventory is to provide a convenient source from which required materials can be obtained in a timely manner. When items are not ordered until needed, the required repair is delayed until delivery, which results in substantial productivity and economic loss, not to mention cost of expedited delivery. Minimization of such losses is the purpose of inventory. The more reactive (as opposed to proactive) that the operating culture and environment are, the more crucial inventory becomes.

C. *Receiving*

Receiving is responsible for physical receipt of delivered items, inspection and count verification thereof, matching shipping and purchase order documents, and delivering the received items according to nature or instructions.

D. *Maintenance Planning*

Planning is responsible for preparing maintenance plans for jobs to be executed efficiently and effectively at a scheduled date. This includes preparation of effective job plans with reliable duration and labor-hour estimates, initiation of purchase order requests and stores requisitions, and reservations upon inventory to meet demands of planned jobs.

E. *Maintenance Supervision*

The contribution of maintenance supervisors and technicians to reliability is effective and efficient execution of scheduled and other requested work with related feedback for continuous improvement.

F. *Inventory Steering Committee*

This committee is comprised of multi-functional representatives from Finance, Purchasing, Stores, Maintenance, Engineering, and Operations. Due to the impact of inventory on working capital, the committee is chaired by the financial function. Other committee members represent perspectives, interests, and insights of the functions they represent.

The committee's mission is to guide the inventory

optimization process. All additions to and deletions from authorized inventory as well as all adjustments of inventory parameters are reviewed and approved by the committee. This applies to items (SKUs), quantities, specifications, etc. No authorized SKU is discontinued or discarded without risk analysis.

Periodically, the committee also reviews trends of material management metrics, and initiates appropriate actions.

G. *Plant/Project Engineering*

Engineering is responsible for protection of asset/process integrity. In that regard, the function champions the management of change process wherein all additions, alterations, modifications, and deletions to facilities, processes, equipment, and other assets require technical approval of Plant Engineering.

Part of this broad responsibility includes configuration management. This includes all documentation that is associated with each asset. Engineering plays a key role in the establishment, collection, assembly, and on-going upkeep of all reference materials associated with equipment and spare parts. This includes history of all associated trials, errors, and successes so that future generations of management preserve the successes and do not repeat the errors and failures.

H. Maintenance/Reliability Engineering

Maintenance/Reliability engineering champions the Reliability Centered Maintenance (RCM) process. This includes the elements of Total Productive Maintenance (TPM) and Root Cause Failure Analysis (RCFA). TPM establishes a maintenance plan for the entire life span of each asset. RCFA is the methodology by which process failures are studied to determine root causes and identify corrective actions to eliminate or mitigate repetition.

- Responsibility for establishing the required level of production capacity rests with asset proprietors.

- Responsibility for design of assets to provide required capacity and to protect it through the management of change process rests with Plant Engineering.

- Determining how best (what, why, how often) to maintain design capacity rests with Maintenance/ Reliability Engineering.

Appendix F (1 and 2) shows the organizational relationships among the eight functions.

Process Steps to Address These Barriers (continued from previous section)

The seminar should present, explain, support, and

sell the principles and concepts commonly associated with world-class maintenance and reliability. A workshop should then be facilitated to help participants tailor local beliefs they agree to mutually share.

3. Establish Shared Beliefs.

See Appendix A.

4. Establish a Reliability Vision Statement.

See Appendix G. Inculcate the statement throughout the organization. This statement succinctly characterizes upper management's ultimate vision for the overall organization.

5. Establish or Refine Functional Mission Statements.

See Appendix H. These statements, which support the overall vision, are developed for each function. They briefly characterize the mission of each organizational unit, functioning in harmony with other units to achieve the ultimate vision.

6. Clarify and Document Functional Roles and Responsibilities.

The general responsibility statements summarized in Obstacle 2 are provided as a starting point. Comprehensive position descriptions, a flow process chart of the work management process (WMP), and RASI charts are also needed. RASI charts clarify Responsible, Accountable, Support, and Information Only functions for each step of the work management process.

7. Develop a Master Plan of Action Steps.

Conduct an assessment (Process Step 1) of the current state of Maintenance/Reliability for comparison to established beliefs and vision. Identify the gap and establish a Master Plan to close it. A generic example is presented in Appendix D, previously introduced.

8. Develop the Budget.
The budget is necessary to implement the plan and build the partnership.

9. Build Justification for Approval.

10. Establish Ultimate Goals and Interim Targets.

These metrics help track progress relative to the master plan. They stem from identification and quantification of the Iceberg and the associated Improvement Potential (Appendices B and C, previously introduced).

11. Establish Incentives.

Each organizational unit should have incentives supportive of overall reliability goals. Clearly identify the required contribution and support necessary from each unit in order to achieve the improvement potential. Establish associated evaluation measures and rewards (financial and other). The incentives for each unit must be harmonious with those of other units. They must not be in conflict. Combined optimal for the overall entity must always be the goal. Group incentives meet this purpose better than individual incentives do. Realization of most targets and goals should reward all contributing parties. Any attempt to quantify the percentage of credit that each functional unit or each individual deserves is counter-productive to inculcation of the partnership.

12. Obtain Management Authorization to Continue.

13. Nurture a Covenantal Partnership among the Functional Organizational Units.

Reward Partnership. Competition should be reserved for external competitors.

14. Sell Maintenance/Reliability Excellence to All Parties throughout the Organization.

15. Implement, Coordinate, Monitor, and Achieve!

OBSTACLE 3
Emphasis on Cost Reduction Rather than Reliability Excellence

A holistic approach to maintenance excellence should be pursued rather than narrow pursuit of mere maintenance cost reduction (see Appendix I). Maintenance cost reduction and operational reliability are frequently opposing initiatives, especially when the existing environment is reactive in nature. Reactive environments commonly result in excessive deferred maintenance. Cost reduction decreases the maintenance budget and results in the deferred hole becoming progressively deeper. When in a dangerous hole, the first rule is: Stop Digging!

Preventive/predictive maintenance is not conceived to put equipment into reliable condition, but to keep it in such condition. Therefore, the essential first essential step to reliability is often to address all deferred workload in order to bootstrap the equipment back to a condition of maintainability and, in turn, reliability. This step involves backlog reduction, overhauls, rebuilds, and replacements. In other words, more maintenance rather than less maintenance! So, when improved reliability is the objective, cost reduction is the wrong managerial decision — at least initially.

Ultimately, maintenance excellence will achieve

operational reliability while also yielding maintenance cost reduction — provided that excellence is pursued first! When cost reduction is pursued first, the reactive hole of deferred maintenance simply gets deeper, in which case, excellence and reliability cannot be achieved.

Process Steps to Address These Barriers (continued from previous section)

16. Purify Current Backlog.

It is likely that some work orders have been completed but never closed out. Others may not be required any longer. Clarify the current status regarding reasons for delay. Capture any work that needs to be performed, but has not been committed to a work order (see Step 17).

17. Aggressively Inspect the Facility and Equipment.

Identify all additional work required to bring equipment and the facility to maintainability.

18. Bootstrap All Assets Back to Reliability.

This step involves working off any deferred backlog to bring it into control limits (4–8 weeks, see Appendix J — Work Program) and bring equipment to a state of maintainability.

19. Reinvigorate the PM/PdM Process.

Conduct it as a controlled experiment until it becomes optimized.

Controlled Experiment — Steps

1. Design experiment.
2. Conduct experiment exactly as designed.
3. Evaluate results.
4. Adjust experiment accordingly.
5. Repeat steps 2–4 repetitively until optimization is found.

This is exactly how the PM/PdM process must be conducted and why 100% PM/PdM schedule compliance is an essential reliability metric.

20. Apply Root Cause Analysis.

Analyze any failures that still occur.

21. Applaud, Honor, Reward, and Preserve Pro-active Results.

OBSTACLE 4

Failure to Effect Cultural Transition from Reaction to Pro-Action, with Adjustment of Expectations

Product quality, on-time delivery, regulatory compliance, optimal profitability, and return-on-asset value are mutually dependent upon reliable operating capacity. Reliability goals are achievable only within a pro-active environment of maintenance excellence, driven by a pro-active culture to which the entire organization is committed. This vision must guide all organizational units in harmonious pursuit of ultimate business objectives. Otherwise, day-to-day efforts are reactively focused on short-term targets of the individual units, to the detriment of long-term goals of the business entity. Compliance with weekly schedules (established to improve reliability) suffers in the interest of near-sighted daily urgencies. All units must understand that scheduled reliability jobs are the important work performed by Maintenance. Responses to failures, although urgent, are not the important work. They usually accomplish little beyond restoration of the asset to unreliable status quo, and are distractions from ultimate reliability goals. Quick fixes seldom eliminate the defects.

When short-term, reactive decisions must be made, don't allow them to diminish your on-going commitment to the ultimate pro-active vision. The struggle for integrity between vision and practice must be immediately restored. Consider what happens when a pro-active reliability work order is reactively bumped in order to respond to a last-minute customer order. By responding to one customer in Week A, how many customers will be disserved in subsequent weeks due to diminution of reliable capacity stemming from disregard of the PM/PdM Process? It was not conducted as a controlled experiment! Management must convey that instant satisfaction is no longer the expectation, except for real emergencies. The emergency priority code has historically been abused.

Process Steps to Address These Barriers (continued from previous section)

22. Drive the Organization to 90+% Weekly Schedule Compliance.

Weekly versus daily schedules is crucial. Compliance to daily schedules delivers a destructive, reactive message as it implies that 24-hour lead time is sufficient. Wrong, because many maintenance jobs cannot be properly prepared overnight. Weekly schedules and weekly schedule compliance are far more indicative of proactive reliability.

23. Realign Organizational Expectations.

The entire organization should understand what expectations they can have for Maintenance response. Except for true emergencies (such as safety, downtime, and environmental situations), prompt response should not be expected. Emphasize regularly that the best way to get all non-emergency work performed is through the weekly scheduling process (Appendix K).

Once the weekly schedule is agreed to, limit the number of individuals with authorization to approve schedule-breaks. Three-to-five senior managers is a reasonable number.

Obstacle 5
Insufficient Organizational Stability to Sustain the Reliability Mission

Cultural change is a gigantic and prolonged challenge. It requires sustained commitment and support of the entire team (all those people who bought into the process and made the commitment to see it through to successful realization and inculcation).

The commitment chain is broken when key members of the committed team move on due to reorganization, promotion, or other reasons. When new managers join the organization, they frequently chose to do things their own way, with their own team. They may come from a reactive environment and may never have been exposed to the benefits pro-active reliability. If so, they are apt to steer a new course even if the essential cultural change is effectively on track. Consequently, the vicious downward spiral of re-active maintenance is reactivated and continues for a number of additional years at the expense of reliability, profitability, and competitiveness.

Process Steps to Address These Barriers (continued from previous section)

24. Strive to Keep the Committed Team Intact.

The team should remain together until the proactive culture is entrenched.

25. Limit Organizational Changes.

Changes could hinder realization of reliability goals.

26. Assure that a Viable Succession Plan is in Place.

27. Establish a Logical and Realistic Master Plan.

This plan should include ultimate goals and interim targets.

28. Apply Trend and Correlation Charts.

These charts should indicate progress toward reliability goals, thereby encouraging managers (original and new) to stay the course.

OBSTACLE **6**
Lack of a Master Plan, Associated Budget, and the Commitment.

Closing the gap between existing reactive conditions and the proactive conditions required to achieve and sustain the level of reliability required to support business success is a major undertaking requiring cultural transition involving many functions of the organization. A master plan of action steps, related responsibilities, and associated time-lines is essential if overall reliability and profitability goals are to be achieved. Master plans provide a road map for all involved parties and highlight the many junctions where two or more steps require close coordination. (See Appendix D again.)

Master plans must be supported with budgets, justifications, and cash flow projections. Industrial engineering or consultant support may be needed to assemble the Case for Approval.

Once approved and funded, implementation of the master plan must be coordinated and monitored. Throughout execution of the plan effective feedback regarding progress is essential. Continued commitment and support must be earned. To this end, a senior multifunction steering committee is recommended. They should designate a singular Champion of the Reliability Process.

*"Setting of Goals is Meaningless
Without Plans for their Achievement"*

Process Steps to Address These Barriers

Some steps previously identified are repeated below and renumbered.

29. Establish a Senior, Multi-function Steering Committee.

30. Develop a Master Plan.

The plan should have action steps with timelines, responsibilities, precedents, and interconnections.

31. Develop the Necessary Budget to Implement the Plan.

32. Build the Justification for Approval.

33. Sell the Plan to All Parties and Obtain Approval.

34. Conduct All Necessary Education and Training.

35. Implement, Coordinate, Monitor, and Achieve.

" It's the culture —stupid!"

CHAPTER TWO
STRUCTURING the MAINTENANCE ORGANIZATION for RELIABILITY

OBSTACLE 7
Maintenance Functions Organized for Reactive Response

Maintenance/Reliability excellence is dependent upon organizational structure fostering pro-action, rather than the reactive structure deployed by most organizations. Unfortunately in these organizations, the bulk of maintenance resources are postured for reactive response to urgencies with insufficient resources remaining for pro-active reliability.

Yet, those same organizations wonder why they live in a constant state of reaction to urgent demands (real or exaggerated). The answer begins with the fact that they have essentially organized for reactive response by decentralizing a disproportional amount of labor resources to fixed-designated assignments within narrow areas of the facility. As a result, technicians spend a significant portion of their day in standby mode, waiting for the next failure to occur. Consequently, few resources remain for performance of important scheduled work aimed at elimination of failures that inhibit asset reliability.

In today's competitive world of downsizing, maintenance labor resources are too sparse to be wasted in standby mode. Deferral of critical maintenance leads to further breakdown, perpetuation of reactive maintenance, and ultimately to failure of the business entity. It perpetuates the reactive status quo, causing the organization's future to look like its past — with continued failure to meet organizational objectives.

To achieve high schedule compliance leading to high process reliability, organizations must recognize and provide for the three broad types of work performed by the Maintenance department:

- Prompt response to urgent demands
- Reliable routine service
  (e.g., PM/PdM procedures)
- Timely relief of jobs in the backlog through effective scheduling

Emergency response is simply the urgent work. It is

not the important work because all it does is perpetuate the status quo. Actually, urgent repair is not maintenance, but the consequence of non-maintenance. Reliable routine service and timely backlog relief are the important work because they improve operational reliability by reducing future failures.

For the above reasons, labor resources must be balanced with workload and distributed organizationally in a manner that assures timely and effective performance of the three broad types of maintenance work, as discussed below.

Work Type Organizational Structure

Two or three organizational structures can be successfully deployed proactively. The key is to find the proper balance between centralized and decentralized deployment of resources. Decentralization provides prompt response and focus of skills on specific assets. It is ideal for emergency response. However, the world-class goal for resources consumed by emergency response is only 10% of total maintenance resources. The remaining resources are best centralized to assure their effective utilization and assignment to proactively-planned and scheduled work (90% of total resources).

Whatever structure you deploy in your own situation, clearing provision for each of the three work types is crucial. The organizational structure discussed below is specifically designed around the three work types and is highly recommended for many situations.

Each work type structure is composed of three major work execution groups (distinct from staff support groups). In turn, each group is responsible for one of the three work type demands placed upon Maintenance. Two minimally-sized groups meet routine and emergency demands (15% and 10% of total resources respectively). The large third group (75% of total resources) is devoted to planned relief of maintenance backlog and major PM/PdM routines (15% of the 75%) requiring more than 30 minutes, shutdown, or a crew larger than two technicians. The nature of each group is explained below. Each requires a different form of control.

A. *The Routine Group Provides Reliable Routine Service*

This group is responsible for the performance of all management-approved routine tasks in accordance with detailed schedules and established quality levels. Routine work is of known content, duration, and frequency. Most prevalent of this workload is the preventive/predictive maintenance process. An optimized PM/PdM process (exclusive of major shutdown routines) requires approximately 15% of maintenance labor resources devoted primarily to inspections and associated corrections (that require only 30 minutes or less). If an inspection uncovers major necessary repairs, such work is normally performed by the backlog relief group, but by the emergency response group when urgency demands. To do otherwise would delay the PM/PdM tour, resulting in non-compliance. Because major shutdown PM/PdM routines require several people to disassemble, reassemble, etc., the backlog relief group performs them. They consume another 15% of maintenance labor resources. Therefore, the total PM/PdM demand on resources is 30%.

Integrity of PM/PdM schedule compliance is protected at all times. Response to emergency requests and scheduled backlog work are not allowed to interrupt the schedule of the routine group. The goal for PM/PdM schedule compliance is 95% or greater (conducted as a controlled experiment).

PM/PdM routines are thoroughly planned when they are established. Further planning is not required until the routine is refined or re-engineered. Through the CMMIS[2], scheduling is automatic as each routine falls due.

---

[2] Throughout this document the acronym CMMIS (Computerized Maintenance Management Information System) is utilized to emphasize that computer support is only an information tool, not a comprehensive maintenance management system as implied by CMMS (Computerized Maintenance Management System). Too many organizations think that CMMS is a complete answer to Reliability Excellence.

B. The Emergency Response Group Provides Prompt Response to True Urgent Needs

This group has responsibility for handling essentially all urgent demands, requesting assistance only when necessary. In meeting these demands, this group protects the other two groups from interruption (approximately 90% of the time). Because the routine crew should never be interrupted, the planned backlog relief group provides the necessary assistance.

The response group is most effective when staffed by multi-skilled technicians. It is radio-dispatched, mobilized, and well equipped (a lean, mean responding team). Urgent work offers little opportunity to be planned or scheduled. Therefore, it places little demand upon planner/scheduler capacity.

C. The Planned Group Provides Timely Relief of All Work Requests Having Adequate Lead Time to be Planned

In a pro-active environment, this group handles the bulk of the maintenance workload (approximately 75%).

Plannable work includes corrective work orders stemming from PM/PdM inspections, major PM routines, small projects (some may be capital), and non-urgent user requests. Once such work has been properly planned and all requirements are at hand, this work is put forth at a weekly coordination meeting as candidates for inclusion on the schedule of work to be performed during the coming week.

As stated above, the planned group is called upon to support the response group (approximately 10% of the time). The other 90% of time, the group works on well-prepared jobs without interruption (schedule breaks). The schedule compliance goal for this group is 90% or higher, which is significantly higher than that of most maintenance departments regardless of industry (60% or less).

Regardless of organizational structure, resources must be deployed so that all three work types are effectively performed in a timely manner.

Process Steps to Address These Barriers (continued from previous chapter)

36. Structurally Organize
Assure performance of all three maintenance work types.

37. Distribute Resources.
Your goal should be approximately 15%/10%/75% (Routine, Response, Planned).

38. Establish Planner/Scheduler Position.

39. Strive for High Schedule Compliance.

The target is 95% for PM/PdM crew and 90% for planned crew.

40. Provide a List of Fill-in Jobs.

The response crew can perform these whenever there is a lull in emergency demands.

41. Trend the Amount of Fill-in Work Performed.

As reliability improves, more fill-in work should be accomplished. This is an indicator that the response crew should be reduced in size.

42. Establish Maintenance/Reliability Engineering Function.

CHAPTER THREE

CLARIFYING REQUIRED ASSET CAPACITY AND ASSOCIATED MAINTENANCE RESOURCES

Capacity utilization associated with invested capital must be optimized. In the competitive world in which we operate, facilities and equipment are increasingly complex and costly. Reliability is indispensable. Return on invested capital is paramount.

OBSTACLE **8**
No Clear Quantification of Asset Capacity Required to Reliably Support Business Plans

Most aspects of reliability are dependent upon maintenance excellence to preserve asset capacity at conditions required to safely produce quality output without damaging the environment. Many operating plans are based upon theoretical gross capacity, with no provision for required performance of pro-active maintenance as required to preserve realistic net Capacity. Without reliable capacity, operating plans and business objectives are unachievable and failure is certain!

Operations, Engineering, and Maintenance should jointly establish the Realistic Net Capacity available to support short-term and long-term business plans. In doing so, they must provide for the essential maintenance routines required to sustain reliable capacity. The difference between theoretical and realistic capacity is dependent upon several variables:

- Deterioration rate of asset reliability (must be provided for when specifying design capacity)
- Interval between restorations
- Skilled labor resource requirement to perform each restoration (measured in crew size and labor-hours required)
- Duration of Restoration, which is the length of time that Operations must plan for equipment to be down when restoration is due
- Design Capacity = Peak Demand (required to satisfy the operating plan) + Provision for Deterioration (between restorations) + Provision for Restoration

Frequency and duration of restoration varies with each asset and the environment in which it is operated. Given greater frequency and/or larger maintenance crew size, less duration is required. The maintenance plan for each asset must be activated when the asset is placed into operation (new, rebuilt, or restored). It is erroneous to assume that PM/PdM can be neglected when equipment is new or recently rebuilt. Activate the routines immediately to keep assets out of the "deferred maintenance hole."

OBSTACLE 9
No Clear Quantification of Maintenance Workload or Downtime Required to Preserve Capacity

To effectively differentiate between theoretical and realistic capacity, a Maintenance Plan must be established for each specific asset operating in its given environment. Plan development is a responsibility of a maintenance/ reliability engineer functioning as champion of the Reliability Centered Maintenance process (RCM). Maintenance plans address the entire life span of a given asset, through pursuit of:

• Maintenance Free Design. Maintenance/ Reliability engineering is the intermediary between project engineering and maintenance. The objective is to assure consideration of maintenance issues and factors during preliminary design, project budgeting, justification, project completion, and final acceptance.

- Preventive/Predictive Maintenance. This area involves origination, monitoring, and continuous improvement of PM/PdM routines in an effort to optimize overall equipment effectiveness (uptime, quality, energy efficiency, yield, and regulatory compliance).

- Maintainability Improvement. This aspect covers identification of equipment modifications and modes of maintenance to minimize loss of capacity and to facilitate ease of maintenance. Root Cause Failure Analysis is a major tool to be deployed.

Operating plans must provide sufficient access to assets for performance of the pro-active routines necessary to sustain reliability, as specified by maintenance plans.

Maintenance plans lead to development and continuous improvement of maintenance routines for each asset in terms of maintenance modes, frequencies, and durations required for preservation of reliable asset capacity.

Maintenance dollars are often spent unwisely. Existing maintenance plans are rudimentary and far from optimized:

• Preventive Maintenance Routines (time-based) may be too many or too few, and performed too often or not often enough.

• Predictive Maintenance (condition-based) may in some cases be more effective than Preventive Maintenance and less intrusive upon run time.

• Sufficient Equipment History may exist to predict life expectancy, leading to just-in-time, programmed replacement with reduction of associated routine inspections, which are expensive. Even when optimized, six inspections might identify only one problem. Thus, five inspections are made needlessly (so to speak).

• Critical equipment receives too little attention. Non-critical equipment with little consequence of failure receives too much attention. In the latter case, the optimal maintenance plan may be to allow the non-critical equipment to run-to-failure. Get every bit of life from it before taking necessary action (replace-ment, rebuild, or repair) because consequences of failure are insignificant.

- Some routines can be performed while equipment is in operation. Other routines require the asset to be out of service. The latter are intrusive upon run time. Such needs must be reflected in quantification of realistic asset capacity.

As these methodologies are optimized, maintenance costs can ultimately be reduced, provided sustained reliability (through Maintenance Excellence) is achieved first.

Process Steps to Address These Barriers (continued from previous section)

43. Project Peak Demand.

Consider the demand to be placed upon each asset by time period (annual, quarterly, monthly, weekly, daily and hourly).

44. Develop Optimal Maintenance Plan for Each Asset.

Set forth the mode, required frequency, and duration of maintenance interventions.

45. Quantify Crew Size, Labor Hours by Skill, and Downtime Duration.

Determine what is required to perform each individual routine and, thereby, what is needed to perform the total Maintenance Plan required for preservation of reliable operational capacity.

46. Establish Realistic Capacity for Each Individual Asset.

Determine the process chains that they comprise. Consider peak demands, maintenance plans, and deterioration rates. Any demand beyond projected peak would exceed design capacity and is not realistically sustainable.

47. Integrate These Needs and Realistic Capacity Determinations.

Develop annual operating and business plans with realistic weekly limitation (e.g., peaks must be limited or realistically provided for).

48. Make Full Use of Valleys in the Annual Operating Plan.

Demand valleys may be seasonal, tied to holidays or to shipping patterns (by day of week or week of month). Maximum proactive use of valleys often yields sufficient equipment reliability to get through operational peaks without maintenance downtime (emergency or planned).

CHAPTER **FOUR**

BALANCING MAINTENANCE RESOURCES WITH WORKLOAD

Thinly-staffed maintenance departments are common in today's world of international competition and associated downsizing. Urgent repair of asset failures and other demands for immediate Maintenance response devour the bulk of available labor resources (manpower).

Urgent tasks (repair of breakdowns) require instant reaction, whereas important tasks (PM/PdM and Planned Backlog Relief) rarely must be performed today, or even this week. The "tyranny of urgency" lies in the distortion of priority. Response to downtime engulfs maintenance resources, but urgent tasks are seldom the most important tasks in regard to long-term objectives. All that urgent tasks do is perpetuate the status quo.

Important work must be distinguished from simply urgent work. To climb out of the "pit of repetitive failures" and achieve reliability, sufficient resources must remain for performance of essential pro-active work that reduces or eliminates repetitive failures, thereby improving future reliability.

OBSTACLE **10**
Imbalance of Maintenance Resources with Maintenance Workload

The process necessary to achieve maintenance excellence and resultant operational reliability begins with macro-planning — the process by which maintenance labor resources are perpetually balanced with maintenance workload in all three basic forms (prompt response to true emergencies, reliable routine services, and timely backlog relief). Macro-planning is distinct from micro-planning, which is development of detailed job plans for the performance of individual maintenance jobs.

Emergencies must be responded to immediately. PM/PdM routines must be reliably performed if potential failures are to be identified in time to preclude them. After providing for these two absolutes, resources must remain for performance of required corrective actions, process improvements, alterations, modifications, and additions. Otherwise, the deferred maintenance pit deepens.

Backlog/Resource Management

Maintenance is classically managed by controlling backlog. Internal customers of the Maintenance function deserve to have work performed on a timely and reliable basis. If this basic expectation is to be fulfilled, after providing for response and routine workload, backlogs must be kept within control limits established in the interest of service and reliability. Backlog is the net known workload; measured in net labor-hours, yet to be completed. If a job has been started, only the portion still to be completed remains in backlog. Backlog does not connote delinquency. Some may be delinquent and some is not delinquent. Backlog is continuously created as:

- Equipment is utilized.
- Processes are modified.
- Facilities are exposed to weather, intended usage, and traffic.
- Inspections identify necessary corrective actions.

Backlog management applies Statistical Process Control (SPC) to the arena of maintenance/reliability.

• Backlogs above the upper control limit turn so slowly that customer needs cannot be met on a timely basis. In addition, emergency work displaces scheduled pro-active jobs, and maintenance is deferred. When legitimate backlog increases beyond upper limits, resources must be increased to protect reliability and bring backlog into acceptable limits.

- Backlogs below the lower control limit contain insufficient workload to fully utilize available labor resources in a smooth, fully scheduled manner. When this occurs, there is tendency to fill the day by stretching job and downtime duration. Resources must be decreased.

Balance must be preserved; otherwise reliability cannot be established and sustained. Resources should be flexed up or down as trends indicate the necessity. There are three resources that can be flexed:

- Overtime is the easiest, assuming that it is not already excessive. An overtime range of 7–15% of paid straight time is recommended. Too little overtime is not an economically sound maintenance strategy. Too much overtime is not healthy, safe, or productive for either company or employees.

- Supplemental contract support is the second most flexible.

- Permanent staff is the least flexible. It should be chosen only when resources and workload are out of specified range in terms of backlog weeks and no change is in sight. The condition appears to be steady-state and permanent.

Two-to-four weeks of "Ready" backlog and four-to-eight weeks of "Total" backlog are commonly accepted control limits. "Ready to Schedule" backlog is part of, but isolated within "Total Backlog."

Excessive backlog leads to unacceptable delay in response to customer needs and ultimately the "D" word — Deferred Maintenance. The "hole" of deferred maintenance gets deeper and deeper. Deterioration gains the upper hand, customer service suffers, and asset deterioration accelerates. The state of reactive maintenance grows asymmetrically. Quickly, the morale and effectiveness of the maintenance organization becomes compromised. This situation is very common and deleterious to reliability, customer service, delivery performance, morale, quality of output, safety, energy conservation, and environmental protection.

When reliability is poor and the environment remains reactive, it is impossible to climb out of the deferred maintenance hole without an influx of resources:

• Lack of customer satisfaction contributes to the low esteem in which the maintenance function is held.

- PM/PdM inspections continue to generate additional backlog by identifying need for corrective actions. Inspections are the investment.
- Without adequate resources, deficiencies found by the inspections are not likely to be corrected in time to avoid failures. Timely corrections are the return on investment. Why invest if sufficient resources are not provided to reap the return?
- People requesting lower priority jobs notice that their requests take forever to be performed. They lose confidence in the system and revert to the practice of coding every request as an emergency.
- Emergency response consumes more and more resources.
- The imbalance between workload and resources spirals out of control.
- Unless Maintenance stays abreast of deterioration, reliability cannot be sustained. The reactive cycle is vicious.
- There are three rules to be applied when an organization is in a deferred maintenance hole:

  - Rule #1 — Stop Digging

  - Rule #2 — Climb Out

  - Rule #3 — Resume the Journey to Reliability

Without measurement, there is no legitimate basis for staffing or for improvement. Accordingly, current backlog upon each crew (measured in weeks) must be calculated and trended on a monthly basis. The vehicle for the calculation is the Maintenance Work Program (Appendix J). This is used to:

- Establish staffing required to keep backlog within established control limits
- Identify necessary adjustments (overtime, contracting, staffing)
- Ensure that expectations for backlog relief are realistic
- Quantify labor-hours of work to be loaded to each crew's weekly schedule (a fair, reliable, yet challenging expectation)
- Assure that even low priority jobs reach the schedule in a reasonable period of time
- Clarify capacity to handle project work and thereby determine when to use contract support
- Provide sufficient resources to achieve reliability

Work programs should not only be calculated for the total Maintenance organization, but also for each maintenance team, skill, shift, and weekday backlog versus weekend backlog. The total calculation provides the average situation, which might indicate sufficient resources. However, individual calculations may indicate poor distribution of the resources relative to where specific workloads lie. The reason the average backlog looks okay may be due to one group being out of control on the high side of control limits, while another group is out of control on the low side. The real situation may be a chronic shortage of specific skill-sets and skill-levels required to address specific workloads essential to asset reliability (see next chapter on skills training). Such imbalance is common and explains why backlog continues to grow further out of control despite apparent adequacy of authorized head-count.

> *"If one foot is on a block of ice and the other is on a hot stove, the average may be just right but both feet are killing you!"*

Process Steps to Address These Barriers (continued from previous section)

49. Provide for the Demands of All Maintenance Plans.

Those plans should be supportive of Reliability.

50. Quantify Workload by Nature,

Workload can be prompt emergency response, reliable routine service, and timely backlog relief. Evaluate the adequacy of labor resources (staffing) to meet the demand. Make allowances for indirect consumption of resources, such as vacations, absence, training, and meetings.

51. Regularly (Quarterly) Balance Resources with Workload.

As appropriate, increase or decrease overtime, contract support, and staffing. The order shown reflects ease of flexibility.

CHAPTER FIVE

SKILLS REQUIRED TO MAINTAIN ADVANCING TECHNOLOGY

There is a crisis within Maintenance organizations — a large and growing shortage of skilled technicians. It is estimated that for every ten technicians retiring today, only three to seven are replacing them. Although the skilled craftsmen developed through historical apprenticeship training processes are retiring, young workers in adequate numbers are not being attracted to a maintenance career.

At the same time, equipment continues to become more complex and sophisticated, requiring much higher skill levels to diagnose and maintain. With installation of technically advanced equipment, a comprehensive skills training process for maintenance technicians and supervisors, as well as operators, becomes increasingly important. This need and the associated training costs are often overlooked when planning and justifying capital projects.

OBSTACLE 11
Skills of Many Maintenance Personnel Fall Short of Levels Required to Maintain Advancing Equipment Technology

This obstacle links with the previous two and applies to incumbents as well as personnel new to the maintenance field. Resources fall short of workload, not only in terms of required headcount but also in terms of required skill level. Therefore, the efforts of Maintenance/ Reliability Engineering to establish a maintenance plan for each asset must incorporate identification of skills and levels of skill required to effectively perform each maintenance plan.

Sources of Skilled Technicians

Readers of this book are either already feeling the effects of maintenance technicians leaving without suitable replacements, or soon will be. When technicians with required skills are obtained, they represent an asset that is too valuable to lose. Invest training dollars in people with proven track records, good work ethic, willing attitude, company loyalty, and great attendance records. They must have the required aptitude and intelligence to learn technical skills.

Some technicians can be found externally:
- Retiring military
- Tech school graduates
- Those laid off from plants that are shutting down

Because these sources will not fill the entire need and will not be promptly integrated into the local organization, the preferred source is promotion from within.

Education and Training Options

There is a critical difference to be understood here. Education can be delivered verbally. Training requires hands-on learning. Highly-technical skills require a combination of education and training. The need is for technicians who can diagnose problems quickly and make effective repairs that eliminate or mitigate frequency of repetition. Too many maintenance personnel are little more than trial and error parts-changers. Note in this case they are referred to as personnel and not by the complimentary term of technician.

There are several developmental modes by which to educate would-be technicians and to upgrade skills of maintenance personnel already on the payroll. The modes each have strengths and weaknesses.

A. *On-the-Job Training (OJT)*

OJT deploys an experienced person to train a new or less experienced person. This approach can be effective provided you have highly-skilled technicians who are willing to share their knowledge and are good instructors.

B. *Vendor Training*

Vendor training can range from excellent to poor and from free to expensive. There is little correlation between cost and worth. You must be careful that the vendor's trainer is not just conducting a sales pitch. Also, the trainees must have the basic education to comprehend what the vendor is teaching. For example, it does no good to send trainees to a DC Drives class if they do not understand basic electricity.

C. *Technical Schools and Community Colleges*

Most parts of the country have good tech schools that are generally low cost. They are an important part of the education/training mix of sources. Selected courses for specific individuals can provide great value (e.g., Welding and Machining). However, in some geographical areas and other circumstances they may not be a complete solution:

- On-campus course scheduling (e.g., sixty minutes, two or three times a week) may not yield essential results soon enough.

- In many organizations, maintenance personnel work significant overtime. For this and other reasons (including family), many are less than enthusiastic about attending night or weekend courses. On-site courses during the normal, paid work day may be necessary.

- Equipment specific skills are often the shortfall. Although some tech and community schools are willing to develop special courses, a minimum of ten attendees is usually required. How many organizations can spare ten maintenance employees at the same time? Sometimes coordination with other area companies can provide the necessary class size.

- Tech schools often have difficulty providing their regular instructors after hours. In such cases they retain, say, retired electrical engineers to teach the course. They may be very knowledgeable regarding the technical education from the book, but lacking the hands on experience necessary to train maintenance personnel in the application of the knowledge.

D. Self-Study

There are excellent self-study materials available regarding maintenance skills, both hard copy and on-line. Although they provide an excellent foundation in basic theory, without accompanying hands-on instruction students fail to gain practical, on-the-job application skills.

E. Professional Educational Organizations

There are a number of organizations that provide hands-on education in concentrated doses, typically weeklong classes with supportive hands-on reinforcement. Although more expensive than government-subsidized tech schools, they often provide better return on the skills training investment. See Appendix L for Sources of Support.

Regardless of the source, optimal education and training processes often deploy a combination of modes, including self-learning (printed texts, computer assisted), video, vendor, simulator, and on-the-job training.

Look into state funding for job education. Many states cover a significant portion of the investment.

The Skills Training Process

Effective training is a process, not merely a list of isolated courses to attend.

Step 1

This step begins with an assessment of skills needed to maintain and operate the specific equipment of which the operating process is comprised. "The process dictates required skills".

Step 2

This step assesses skills possessed by incumbent personnel relative to required skills.

Step 3

Any gap between the two defines training needs in terms of courses to be provided.

Step 4

Develop a progressive advancement/reward scheme to entice existing and new maintenance employees to enter, commit to, and remain with the learning opportunity. Once an organization invests in development of high-skilled personnel to protect major capital investments, it cannot afford to lose the resultant, highly-skilled people to competitive employers. Therefore, integrate achievement of skills with wage administration.

Personnel bidding into maintenance should be tested for necessary aptitude. External personnel applying for a vacancy that cannot be filled internally should be tested to determine what skills they truly possess and to determine their appropriate point of entry within the established pay scale and within the training process. To qualify for new pay grades associated with advanced skills, incumbent personnel must participate in the training process. Those with significant seniority choosing not to participate in the training program are red-lined at their present pay level. The only increases they will receive are cost-of-living increases. Written tests as well as application tests should be administered before acceptance into the Maintenance organization and before pay grade advances are awarded.

> *"Many common problems are effectively addressed by pay-for-skills".*

Pay for Skills

The most cost-effective programs for educating would-be maintenance technicians and upgrading existing technicians contain pay-for-skill concepts. Such programs provide financial incentive for employees to maximize their learning opportunity and to effectively apply their learning to the work place. Typical process provisions include:

- Testing at conclusion of each class
- Wage advancement is denied if courses are failed and the courses must be repeated
- Advancement, promotion, and increased pay rate for successful completion of blocks of courses
- Minimum and maximum time limits at each skill level
- Cross-training to expand multi-craft skills

Although Human Resource functions might resist, pay-for-skills should be given serious consideration. See Appendix L for sources of further information.

Process Steps to Address These Barriers (continued from previous section)

52. *Conduct a Needs Assessment.*

Determine the skills required to establish and preserve reliability of the process.

53. *Identify Incumbent Skills.*

Determine existence of these skills by individual.

54. *Clarify the Gap.*

Identify required Education and Training Course Matter.

55. Develop a Progression Requirements and Reward Model.

56. Determine How the Skills Crisis Will Affect Your Organization.

Develop a succession plan to address the problem.

57. For Each Technician, Identify the Proper Entry Point Into the Training Curriculum.

Base your analysis on the skills gap of each individual. Incumbents are at difficult levels in relation to skills required to maintain the process. New employees may or may not bring some of the required skills.

CHAPTER **SIX**

MATERIALS SUPPORT

To sustain a cost-effective, responsive, and reliable operation, the maintenance department is dependent upon reliable and timely logistical support. In reactive operations, twenty-five percent of equipment downtime duration is due to lack of spare parts, materials, supplies, and special tools necessary to perform repairs correctly. Nothing is more frustrating to maintenance technicians than to start a job only to find that it cannot be completed due to lack of necessary spare parts or other required materials. Downtime is exponentially extended when required parts must be sourced, procured, and received. Lack of effective logistical support results in reliability problems with economic consequences:

- Magnitude of disruption is not always proportional to the cost of the missing part. The required item may be an expensive part or component unique to a specific equipment unit, or it may be a common hardware item costing less than a few dollars.

- On the low end there may be loss of maintenance pro ductivity as resources are shifted from an unfinished job to an urgent situation (real or exaggerated).

- On the high end, the consequence might measure in thousands of dollars due to major disruption of quality output and resultant delinquency of customer deliveries.

Regardless, it is critical that the item be readily available or quickly be made available so mechanics are able to complete the repair promptly and properly. Optimal reliability cannot be achieved unless spare parts and other materials are promptly available when needed.

OBSTACLE **12**
Procurement is Focused Too Much on Initial Cost vs. Life Cycle Cost

This obstacle applies to equipment, spare parts and maintenance materials.

The maintenance storeroom is often characterized as A Productivity Booby Trap! The following complaints are common when purchase price and inventory value are minimized rather than optimized:

- "Purchasing buys on cost alone!" Reliability is deni grated when procurement is based on lowest bid. The consequence of less reliable spare parts costs Production multiples of what Purchasing saves. When goals of one functional unit of any organization are maximized, goals of the business entity become sub-optimized. John Ruskin, an early economist, character-ized this dichotomy in the following manner:

The Lowest Bid

"It is unwise to pay too much, but it is worse to pay too little. When you pay too much, you lose a little money — that is all. When you pay too little, you some-times lose everything, because the thing you bought was incapable of doing what it was bought to do (reliably[3]). Common law of business balance prohibits paying a lit-tle and getting a lot — it cannot be done. If you deal with the lowest bidder, it is well to add something for the risk that you run. If you do that, you have enough to pay for something better."

[3] Author's clarification relative to current subject

- "We are tired of Band-Aid repairs." Reliability requires jobs to be done properly on the first attempt.

- The storeroom regularly runs out of spare parts and

other materials that are urgently needed to keep critical equipment on-line.

- With no apparent awareness of periodic criticality and without consulting Maintenance or Engineering, materials management unilaterally discontinues stocking seldom-used spare parts, without adequate consideration of criticality when they are needed.

- When direct purchases are received for specific jobs, the originators of the request are not informed. Nor are they alerted when delivery will be delayed.

- Maintenance technicians must go to the storeroom in order to identify required parts. If a stores catalog exists, it is ineffective.

On the other hand, maintenance-related inventory is a major portion of working capital. Therefore, management of inventory must be optimized based on financial impact upon the overall business entity:

- Get the right materials to the right place, at the right time, at optimal cost. Adequate inventories must be maintained to minimize delays associated with stock outages and, in turn, to protect the plant against the related extension of downtime.

- Avoid excessive inventories. Inventories must be designed and controlled so that carrying costs do not exceed costs associated with extension of downtime.

- Although the two previous objectives appear to be in conflict, their optimization lies in engineering economics. Each stocking decision must reflect the lowest overall cost to the operation. Unit cost of delivered material, cost of carrying inventory, and asset criticality (the consequence of not having the item on hand when needed) vary greatly by item (Stock Keeping Unit or SKU).

Often there is debate regarding advantages and disadvantages of Just-In-Time (JIT) acquisition versus Statistical Inventory Control (SIC). Both have application to the inventory optimization process leading to sustained reliability.

A basic prerequisite of JIT acquisition is a proactive culture and environment. Project work and planned maintenance jobs can rely on JIT because lead-time is available and jobs should not be scheduled until all required materials are on-hand. However, emergency breakdowns give no lead-time. Therefore, when an organization is in reactive mode, adequate inventory must be on hand. Otherwise, downtime increases, output and quality decrease, and customer service and profitability suffer.

The combination of JIT for proactive work and SIC supportive of reactive work must be optimized. The approach to be taken is JIT, provided the need is not immediate, the job can be scheduled, and needed materials do not involve extended delivery time. For urgent needs, inventory is necessary unless a local, reliable, and prompt supplier can be identified and established as a working partner. Such a relationship might include a consignment inventory agreement.

Associated Benefits

- Proactive organizations are less dependent upon inventory. Early PM/PdM warning of developing conditions yields the necessary lead-time to plan and procure using just-in-time concepts and procedures.

- Once Maintenance/Reliability excellence and a proactive environment have been achieved, emergency response should represent no more than 20% of maintenance work orders and consume only 10% of maintenance labor resources. The different percentages occur because emergency response jobs average significantly fewer labor hours than planned jobs and projects.

- Parts and material cost for small, emergency jobs is also less than for large, planned jobs. Assuming the environment is proactive, no more than 25% of maintenance material cost need be acquired from store room inventory. At least 75% should be acquired just-in-time. When the environment is still reactive, the redundancy of emergency jobs detracts significantly from this advantage; 50% or more of material cost is acquired through storeroom inventory.

Associated Prerequisites

- All functional units must effectively meet their reliability-related responsibilities. Purchasing, Stores, Maintenance, Engineering, and Operations constitute one of the most important partnerships within the operational arena. They share responsibility for materials support and control.

- Purchasing provides a responsive process for timely procurement of all parts and materials directly purchased or authorized to be held in inventory. Procurement includes sourcing, purchasing, and expediting.

- Stores maintains a reliable inventory (on-premise or near-by) to meet reactive demands (emergency repairs) when equipment fails; to support routine maintenance efforts designed to avoid future failures; and to assure availability of insurance spares. This requirement necessitates a well-stocked and well-controlled in-house storeroom or alternate source, containing:

- Frequently-used parts and materials that can be purchased less expensively in bulk then in minor quantities.
  - Less frequently-used parts that are either critical to operation of plant and equipment or subject to long delivery times.

- Receiving is responsible for physical receipt of deliveries and associated information processing:
  - Matches shipping documents to receivers and visually inspects received items for conformity, count and damage
  - Delivers stock replenishments to storeroom
  - Relates direct purchase items to cross-referenced work orders and stages them in a planned work order cage
  - Notifies Planning when all direct purchase items for a given work order have been received, or, if requested, notifies Planning as each direct purchase item has been received

- Vendor relationships are effective and reliable, and can be characterized as Partnerships.

- Maintenance provides Purchasing with sufficient lead time on all proactive work to enable Just-In-Time (JIT) procurement of required materials.

- The entire organization understands and honors the purpose of maintenance stores:
  - Storeroom inventories are primarily for emergency needs.
  - Inventory is not depleted to support projects that by nature can be planned. For example, stock levels as designed may be sufficient to support a demand of four units per week with a lead time of two weeks (Maximum Inventory = 8). If six are pulled for a project and an equipment failure occurs requiring the same SKU, Production will likely suffer extended downtime until the associated stock can be replenished with the expense of rush shipping.

- Equipment History effectively supports pursuit of reliability. The cost of all parts and materials is charged to appropriate work orders, and thereby to appropriate equipment units, cost centers, and accounts. This requires inventory accounts wherein the value of spare parts is carried until time of issue.

  Should Accounting choose to expense materials and spare parts at the time of receipt, the value of each SKU must be maintained within the CMMIS materials module so that material costs can be reflected in equipment history at time of usage.

Vendor Partnerships

A few thoughts are offered here regarding optimization of procurement supportive of sustained reliability and ultimate goals of the enterprise. Reference has already been made to the common corporate climate wherein extreme pressure is placed on each business function to capture all possible cost reduction. In response, each function applies its own expertise to its own sphere of influence, as encouraged by functional incentives. Consequently, industry is on a management treadmill. The search must be for optimum in all situations.

> *"Whenever individual functions maximize their own interests, the entity is sub-optimized."*

There are four material management initiatives commonly associated with world-class reliability:

- National contracts to capitalize on the purchasing power of the corporation. The corporation is able to negotiate better pricing than individual sites can negotiate.
- Single source supply to minimize administrative burden upon purchasing and accounts payable.
- Partnership with local sources that provide exceptional service and technical support on engineered products (as distinct from commodity products).

• Contract storeroom management to extricate internal resources from non-core activities.

Pitfalls

Individually each of the four initiatives offers benefit. However, world-class operations search for an optimized hybrid of the four. Pitfalls of singular application of each initiative are explored below:

• *National Contracts with Single Source Suppliers*

Often the top two initiatives are merged into a single endeavor and serve the singular interests of Procurement quite well, but sometimes to the detriment of the overall enterprise. National contracts are frequently horizontal, single-source agreements, whereby a national supplier expands its product line to include the broad range of parts, components, assemblies, and materials required to support facility construction and plant maintenance — thereby, becoming a single source supplier.

Such agreements can be far from the optimized procurement process sought. They often lack the critical service, response, and technical support necessary to sustain reliability and to support a just-in-time culture:

• National contracts are most appropriate for commodity-type hardware items, such as standard fasteners, belts, and hoses. Unlike engineered parts, commodity items are less critical to reliable operation, safety of personnel, and protection of the surrounding environment. They also require less time to acquire.

- National contracts are often awarded on comparison of the national suppliers' list price versus the local suppliers' delivered price. The unasked question is where will the national supplier deliver from? If it has a nationwide distribution network, lower list price may yield lower delivered price. Otherwise, shipping and handling charges on frequent, small, and rush orders will exceed list price savings.

- How can a national supplier, without a broad network of distribution centers, respond to:
  - Midnight calls for 7:00 AM delivery?
  - Prompt, on-site technical support? It is unlikely that any single-source suppliers will have technical product knowledge across their (expanded) array of products. Such technical knowledge is critical in the case of engineered parts and materials (as opposed to commodity items) and should be available from a true partner/supplier/technical advisor.

- *Partnership with Local Sources*

    When making cost comparisons with national vendors, the value provided by relationships with specialty distributors must not be undervalued when striving for an optimized procurement strategy.

Horizontal, national, single-source agreements weaken the position of local, specialized suppliers and jeopardize the many value-added services that they offer and which individual sites depend upon, especially in emergency situations. To the chagrin of local managers responsible for reliable output, their partnerships with local sources are often decimated when single-source national contracts become a corporate edict.

Elimination of some local vendor relationships may be a good thing. But, preservation of other local vendors can be critical. In many cases, the nature of required support is most effectively nurtured with high-quality local distributors, who are best postured to:

- Respond to emergency needs (24/7, even at 3:00 AM if needed).
- Help evaluate failures through specialized knowledge, and bring factory representatives to bear as needed.
- Keep local buyers, managers, engineers, project managers, and maintenance technicians abreast of the constant stream of technical advancements, new products, and other enhancements.
- Locally train the same parties in product application.
- Help source hard-to-find items.

Local vendors cannot be expected to respond to emergency situations without the opportunity to supply a portion of normal daily requirements. Otherwise, the prices they must charge for emergency support (24-hour, 7-day-a-week service, including holidays) will drastically diminish list price savings associated with standard delivery terms of many national agreements.

- *Contract Storeroom Management*

The fourth pitfall concerns contract storeroom management. In addition to managing the storeroom, the contract manager may also be responsible for procurement of stock replenishments and possibly for procurement of direct purchase items (not authorized to be carried in inventory).

This approach can be part of an optimized procurement strategy, provided the contractor is carefully selected and a clear, detailed contract is established. The contractor should:

- Own the stock until requisitioned
- Manage all storeroom activity
- Process all transactions
- Release replenishment needs against system and blanket purchase orders, previously established by an independent, internal procurement body.

Due to potential conflicts of interest, it is inadvisable for the contractor to independently procure from its parent company:

- Contractors have access to all current suppliers and their prices. With this knowledge, they have an unfair bidding advantage. It is unlikely that replenishment prices will be competitive.
- Likewise, contractors should not establish specifications. Their specifications may be advantageous to product lines of their parent company or affiliates.

The Answer

Part of the answer lies in vertical (rather than horizontal) national agreements. Such agreements are made directly with manufactures within specific product groups (e.g., hoses). Chosen manufacturers should be highly regarded for time-tested quality. Local buyers, engineers, and maintenance managers should participate in the selection, as they know the criticality (safety, environmental, quality, and reliability) of applications and past performance of parts produced by various manufacturers. Pricing should be established nationally by Corporate Purchasing, directly with chosen manufacturers.

Local buyers, with maintenance and engineering input, should select distributors. They know which local distributors are most reliable and have demonstrated a spirit of partnership. If the chosen manufacturer wants the corporate-wide business, it will license and set up the locally specified distributors to carry its product lines.

More than one manufacturer might be chosen within a core product group — those products with natural relationships. However, the selected local distributor becomes the point of single-source for the entire core product group. With this structure, purchasing power of the corporation is captured while unparalleled support of local distributors is preserved. The resultant reliability economies greatly exceed any cost increase associated with processing of payables.

Process Steps to Address These Barriers (continued from previous section)

58. Establish Strong Functions for Maintenance Buying, Storeroom Management, and Receiving.

These functions should effectively support continuous reliability.

59. Establish a Well-Designed and Effectively Located Storeroom.

Include good security and a good location system. By deploying technology, around the clock attendance should not be necessary.

60. Establish an Inventory Steering Committee.

61. Integrate Material Control and Maintenance Work Control Systems.

62. Fully Load the Material Database.

63. Adhere to Effective Material/Maintenance Procedures.

64. Continuously Refine Material Control Parameters.

65. Establish Inventory Reservation for Planned Jobs.

66. Establish Material Deliver Procedures.

Determine how to deliver material to maintenance job sites.

67. Establish Effective Vendor Partnerships.

CHAPTER SEVEN

PREPARATION OF MAINTENANCE WORK

Thorough preparation is essential to safe, effective, and efficient execution of maintenance jobs with environmental vigilance. Striving for timely response to the proactive needs of internal and external customers, it enables realization of asset reliability supportive of stable, long-term and short-term operational output.

OBSTACLE 13
Insufficient Preparation for the Execution of Maintenance Work

Work preparation encompasses job planning, parts acquisition, work measurement, coordination between Operations and Maintenance, and scheduling of maintenance resources to establish expectation for weekly job completions.

Objectives and Benefits of Work Preparation

When preparation is effectively performed by a professional staff organization, maintenance supervisors are relieved of many misplaced responsibilities and are thereby able to devote their full effort to effective direction of job execution and development of maintenance personnel. Craftsmen become more productive because potential delays and conflicts are identified and, if avoidable, resolved proactively before jobs are scheduled. Thereby, morale and job satisfaction are improved. More work is accomplished more efficiently, at lower cost, with fewer disturbances of operations, higher quality (reduced variability of processes), increased longevity of equipment, and reduced parts usage.

• Safety and environmental protection are ensured when well-defined job steps are followed. When flying, prior to takeoff, we certainly want the pilot and co-pilot to go through their checklist. In like manner, those people concerned about loss prevention, property damage, and business interruption want assurance that personnel are following prescribed procedures when performing maintenance work.

• Each hour of effective preparation typically returns three-to-five hours in mechanic labor hours to execute (or equivalent savings).

- Well-prepared jobs require only half as much down time to execute compared to unplanned jobs. The result is improved reliability (measured in uptime hours).

- Equipment availability is optimized (measured in on-line hours).

- Operational interruptions, downtime, and loss of production capacity are minimized.

- Maintenance work is completed more promptly and more effectively, when needed, in a safe and environmentally conscious manner, at optimal cost.

- Asset reliability and external customer service are dramatically improved.

The benefits accrue from the combination of planning, parts acquisition, work-measurement, coordination and scheduling. Additional benefits of these five individual components of preparation are individually discussed below.

Prerequisites of Effective Work Preparation

A number of prerequisites are required if the work preparation effort is to yield optimal results:

- Lead-time is essential. Without it, no time exists to prepare for effective and efficient work execution. Potential problems and needs must be identified as early as possible, with appropriate initiation of work orders by all parties (Production, Maintenance, Engineering, etc.). As a result, backlog of work is known and jobs can be effectively planned, coordinated, and scheduled in preparation for execution at a future date.

- Maintenance and internal partner/customers (users of maintenance service) jointly share responsibility for the cost of maintenance. Users influence the volume and timing of maintenance work. Maintenance influences effectiveness and efficiency of performance. Budgetary practices should reflect the shared responsibility.

Accountability for Maintenance Budget Variances					
		Responsibility			
		Production		Maintenance	
Cost Center	Budget	Standard	Volume Variance	Actual	Performance Variance
A	$600,000	$700,000	$100,000	$750,000	$50,000
B	$400,000	$350,000	($50,000)	$375,000	$25,000
Total	$1,000,000	$1,050,000	$50,000	$1,125,000	$75,000

- Involving cost center managers in establishing the
  magnitude of their share of the overall maintenance
  budget and holding them partially accountable for control of that budget significantly helps to pull them into
  the Maintenance, Operations, Reliability partnership.
  They become focused on investing their maintenance
  dollars in proactive reliability, rather than spending
  them on excessive and often needless, reactive desires.
  - Developed through the RCM Process, a maintenance
    plan must be established for each asset.
  - Resources must be continually balanced with work-
    load.
  - Craftsmanship must be a matter of pride to all main-
    tenance personnel, including:

    - Rebuild and overhaul skills
    - Workmanship
    - Effective use of instrumentation and other
      technology
    - Reliable equipment inspections
    - Troubleshooting logic

- The selected CMMIS must be comprehensively loaded
  and effectively applied.
- PM/PdM inspections must be comprehensive and
  thorough.
- Each equipment failure analyzed using RCFA.
- Equipment history complete and well documented.

- Feedback (via the CMMIS on completed jobs from supervisors and technicians) must be meaningful to equipment history and to refinement of job estimates and planned job packages.
- A comprehensive, up-to-date, and well-indexed (cross-referenced) maintenance technical library must be established.

Planning (How to Best Perform the Job)

Planning is the development of a detailed plan by which to achieve an end (e.g., a maintenance repair or rebuild). The plan encompasses preparation, execution, and startup efforts (pre-shutdown, internal to shutdown, and post-shutdown).

Planning is advanced preparation for each specific job so it can be performed in an efficient and effective manner with the least possible amount of interruption, delay, and downtime.

- It enables procurement and coordination of all logistics necessary for execution of proactive maintenance jobs at a future date.

- It involves detailed analysis to determine and describe work to be performed, the sequence of associated tasks, methods to deploy, and required resources (including skills, crew size, labor-hours, parts, materials, special tools, and equipment), plus an estimate of duration time and total cost.

- It includes identification of safety precautions, required permits, communication requirements, and reference documents such as drawings and wiring diagrams.

- It initiates required procurements using purchase order requests and stock requisitions for parts, materials, tools, and equipment.

- It also includes work measurement to establish job duration and labor hour requirements (see following section on Work Measurement).

Associated Benefits

- Originator expectations and scope are clearly established.
- All resource needs are identified.
- Job instructions are documented.
- Safety and environmental issues are foreseen.

Associated Prerequisites

- An extensive library of planned job packages
- An extensive library of labor estimates for given jobs
- An extensive library of established procedures (e.g., safety procedures)

Process Steps to Address These Barriers (continued from previous section)

68. Establish a Strong Planning Function.

Planners should be full time, and experienced or well trained.

69. Capture All Planning Efforts in Reference Libraries.

This effort will facilitate the ease, accuracy, and consistency of subsequent job plans.

70. Continuously Refine Job Plans.

71. Assure Effective Feedback to Planners from Technicians and Supervisors.

Work Measurement (Job Estimating)

Estimating is an essential building block within the planning process. Backlog management, job scheduling, and efficiency determination all depend upon reliable estimates of duration hours and labor-hours required to perform individual work orders. Work measurement supports:

- Quantification of backlog measured in labor-hours.

- Establishment of meaningful maintenance schedules representing management's expectation as to what is to be accomplished by maintenance resources to be paid during the schedule week. Job schedules are defined in duration-hours. Duration-hours multiplied by crew size equals labor-hours.

- Meaningful schedules translate into realistic promises to internal customers. Promises include when the unit will be taken out of service, how long it will be out of service, and when it will be returned to service.

- The scheduling time-line of individual jobs provides supervision with expectancies by which to evaluate progress throughout execution of each job.

- Establishment of realistic expectancies on which to evaluate supervisory effectiveness and crew efficiency. All personnel work more effectively when evaluated against fair, yet challenging standards whereby they know what is expected of them and if they are performing well.

The two precision levels of work measurement are standards and estimates. For purposes of this discussion, estimates and historical averages are comparable.

Standards quantify how long jobs should take. Estimates or historical averages quantify how long jobs will probably take, given current levels of efficiency. The difference is most significant when maintenance is still reactive and efficiency is, therefore, no higher than 50%. If standard is used (when in the reactive state) to:

- Quantify backlog, then weeks of backlog are seriously understated.
- Establish schedules, then compliance is extremely low.
- Establish return to service promises, then promises are seriously missed.

If estimates are used to calculate efficiency, the result will not be meaningful because it will always approximate 100% — the numerator and denominator will be essentially equal.

The need is for job estimates that establish challenging yet fair expectancies, while feeding realistic backlog information and achievable schedules. Although it is desirable to use management tools with realistic pull to drive improvement, it is not realistic to expect efficiency to improve from 50% to 100% (or even 85%) in the near-term. In fact, 10% improvement per quarter is a lofty goal.

The key is to use standards and measure current efficiency levels. These two parameters can be deployed to temper expectancy, according to which of the three applications the estimate will be applied and how much pull is realistic.

- Backlog Weeks. These should be calculated on current efficiency adjusted by realistic near term improvement (pull).

- Schedules. People work at a pace that is based partially on the amount of work they are given. Ideally, jobs are quantified within the schedule based on standard labor-hours (expectancy as to how long a job should take). However, there is also a customer service aspect to the scheduling process. Because the schedule represents promises and commitments to the internal customer, there is need to accurately predict when a unit will be removed from and returned to service. Therefore, it is crucial that the schedule be attainable, given current efficiency levels. Consequently, schedules are often developed based upon a 10% improvement (pull) over current efficiency. This approach makes schedules achievable without accepting status quo.

- Efficiency. This element compares actual labor-hours charged each job to standard labor-hours for that job (how long the job should take to perform, assuming effective job preparation and a properly-trained techni cian, working at a normal pace).
- Efficiency = Standard Labor Hours / Actual Labor Hours

Measurement Applications

The table below reflects the most appropriate of the two measurement forms to each of the three applications.

Form of Measurement	Application		
	Backlog	Scheduling	Efficiency
Current Average / 1.1 *	•		•*
Standard			•

* The 1.1 divisor puts 10% "pull" into schedules

Associated Prerequisites

Averages and estimates should be continuously refined until worthy of being termed standards. Initially, they likely will not reflect how many labor-hours the job should require (100% efficiency). Continually refine them based upon feedback. Every week, each planner (who-ever is making the estimates) should receive a Top-Ten Report listing the work orders on which actual hours charged differed from estimate by the greatest percentage (plus or minus). Those ten jobs should be reviewed with the supervisor, technician, or originator. Do not automati-cally assume that the estimate needs to be adjusted. The estimate may be good, but performance was either poor or possibly exceptional. Perhaps something was inadequately provided for in the job plan. In that case, improve the plan; do not change the estimate. Should it be concluded that the estimate is wrong, change it! With continual refinement, estimates ultimately become true standards.

Process Steps to Address These Barriers (continued from previous section)

72. Continually Refine Job Estimates and Labor Libraries.

CHAPTER EIGHT

COORDINATION, SCHEDULING & SUPERVISED EXECUTION

Coordination and scheduling are also part of preparation. They specifically address the expectation of work to be completed during the forthcoming week. Without a coordinated, agreed upon, and detailed schedule, the expectation has not been established. All personnel accomplish more when working to a challenging, yet realistic, expectation. Failure to establish expectation reflects abdication of basic managerial responsibility.

OBSTACLE **14**
Insufficient Statement of Expectations

The statement should be agreed upon, realistic, fair and challenging.

The coordination process is a weekly cooperative effort between Operations (custodians of the assets) and Maintenance (providers of asset reliability). Considering near-term demand for output as well as ongoing reliability, the partners jointly achieve consensus as to the most critical work to be performed during the coming week, within the limits of available maintenance resources. Operations is concerned about targets for daily and weekly output (a near-term horizon), whereas Maintenance is concerned about optimized and stable capacity to meet on-going output demands (a long-term horizon). The weekly coordination meeting is the forum by which the near-term / long-term dichotomy is continually balanced.

Coordination further encompasses logistical efforts to assemble and synchronize all resources required to perform each job. Maintenance Planning, Purchasing, Receiving, and Stores all contribute to the assembly effort. Only when assembly is complete, does the job become Ready to be Scheduled (as a status code). During the coordination meeting, all ready jobs are reviewed to select the most important for the approaching schedule week. The resultant list is limited by available resources and operational feasibility regarding release of involved assets for performance of each individual job. The list is force-ranked by importance and the parties agree upon a day and time when assets can be released and maintenance resources can be made available. The agreement becomes a contract between the parties. When schedule slippage occurs (non-compliance), the least important jobs should be sacrificed, not the most important.

Personnel come to realize that the best way to get work properly performed in timely fashion is through the formal planning, coordination, and scheduling process:

- Labor resources, parts, and materials are delivered to the job site as needed, enabling Maintenance to per form work properly on the first attempt.
- Jobs are scheduled within the agreed upon time frame.
- Even low-priority jobs remain in the focused queue (backlog) and are accomplished within an appropriate time frame.
- Asset reliability and customer-service are dramatically improved.
- Operational interruptions and downtime are minimized.
    - Process quality is sustained.
    - Equipment availability is optimized.

- Each hour of effective preparation typically returns three-to-five hours in mechanic labor hours to execute (or equivalent savings).
- Well-prepared jobs, compared to unplanned jobs, require only half as much downtime to execute, there by increasing realized operational capacity.

Process Steps to Address These Barriers (continued from previous section)

73. Designate Primary Liaisons within Operations for Coordination with Maintenance.

74. Carefully Select the Most Logical Schedule Week.

Many maintenance organizations select Friday through Thursday because the greatest equipment access occurs over the weekend and Friday is used to fabricate in preparation for weekend jobs.

75. Determine and Firmly Establish when the Coordination Meeting Should Occur.

Insight to next week's operating schedule must be available, and time must remain after the meeting to detail and post the schedule.

76. Establish Attendees for the Weekly Coordination Meeting.

These should include mandatory attendees, backup attendees, and optional attendees.

Scheduling (When to Do the Job)

Scheduling is the process by which all maintenance activity is activated. It is viewed as the marketing arm of maintenance excellence, as it yields the earliest tangible results visible to Operations — often within weeks. Production grows impatient unless they can see interim progress accruing from scheduling (increased utilization of existing resources.

By contrast, preventive maintenance and job planning require months of focused effort before yielding measurable results.

Scheduling begins when jobs have been planned, material availability has been assured, and coordination with internal-customers has taken place. It is the documented process whereby labor resources and support equipment are allocated to specific jobs at specific times when Operations can release associated equipment. The resulting schedule provides the expectation of jobs to be accomplished during the coming week, given the maintenance resources that are expected to be available.

Without an effective schedule, there is no expectation as to the benefits to be gained from next week's Maintenance payroll. All organizations and all individuals perform better and accomplish more when working to accomplish clearly established, communicated, and published expectations that are fair, yet challenging. Without a proactive expectation, the reactive demands that typically arise are the only work performed. As a result, asset failures devour the bulk of available resources. Consequently, crucial proactive maintenance routines are not performed when due. Reliable output, quality, customer service, safety, energy conservation, and environmental protection suffer the consequences.

Thinking that we work best under pressure, we subconsciously procrastinate until pressure forces us into reaction (management by crisis). Response may be faster, but work performed is less effective and less reliable, due to lack of preparation followed by hasty judgments. Doing the wrong things faster accomplishes nothing. Tyranny of urgency lies in distortion of priority. Breakdowns require instant reactive repair. Preventive and predictive inspections and planned backlog relief rarely must be done today. Although they can be engulfing, the urgent tasks are seldom the most important tasks when viewed from the perspective of business objectives.

Urgency alone (without due consideration to importance) cannot be allowed to determine how vital resources are consumed. Are they to be deployed effectively through scheduling, or ineffectively consumed by reaction to repetitive breakdowns? Either management schedules vital maintenance resources, or equipment failure and impatient customers dictate how resources are consumed and wasted on the wrong jobs. "Prompt performance of the wrong work does not yield reliability."

Schedule Compliance

Operational control of maintenance, though much slower reacting, is similar to the thermostat principal. The fundamental requirement is an objective (the schedule) against which to control, followed by action (execution of the schedule) to achieve the objective. The resultant achievement, measured against the original intention (schedule compliance), provides feedback by which to address deviations.

> *"Schedule precisely, proceed positively,
> and persistently pursue weekly goals."*

Without effective scheduling and high schedule compliance, maintenance is haphazard, costly, and ineffective. Promises regularly fail to be met, thereby causing constant problems for Operations. If Maintenance consistently fails to put equipment back into service as promised, Operations becomes increasingly reluctant to release equipment in the future, at the jeopardy of sustained reliability.

Scheduling Benefits

- The schedule establishes expectancy as to how much specific work is to be accomplished during a given week in return for the payroll investment. Jobs are scheduled in order of agreed upon importance, by day and by individual(s) to be assigned.

- Expectations and commitments are realistically based upon quantity of resources available for pro-active work during the schedule-week.

- Promises made between Operations and Maintenance are dependable as to: when equipment will be taken out of service and turned over to Maintenance, the duration of the Maintenance effort, and when equipment will be returned to service.

- In turn, production control and shipping have reliable information on which to plan and schedule their own activities.

- Maintenance supervisors then have a mechanism by which to:
    - Control work execution
    - Evaluate crew performance
    - Motivate employees who now know what is expected of them and whether or not they are meeting the challenge.

- Management has a vehicle by which to evaluate supervisor performance and utilization of resources. Output and productivity are favorably impacted.

- All identified preventive and predictive routines are performed when due.

- All identified corrective maintenance is promptly performed, thereby precluding future equipment failures.

Process Steps to Address These Barriers (continued from previous section)

77. Conduct PM/PdM and Corrective Maintenance Process as a Controlled Experiment.

78. Perform Corrective Maintenance in Timely Manner.

Otherwise Return from the PM/PdM Investment is Lost.

79. Fully Load All Maintenance Personnel to the Weekly Schedule.

Detail the schedule by job, by day, and by individual to be assigned. See Appendix K for recommended format.

80. Schedule the Least Important Jobs to the Best Troubleshooters.

These are the jobs to be sacrificed whenever true emergencies must break the schedule, and it is the best troubleshooters who will respond.

81. Maintain Concentration on Scheduled Jobs.

Focus on the most important.

82. Identify and Address Reasons for Non-Compliance.

83. Study Recurring Windows of Opportunity.

Make effective use of all available access to equipment.

84. Measure and Trend Crew Efficiency.

85. Persevere—Don't Quit!

Achieve the highest possible level of schedule compliance despite interruptions.

Considered together, the five activities that are discussed above and summarized below constitute Job Preparation.

- Job planning
- Parts and materials acquisition
- Work measurement
- Coordination
- Scheduling

They are supportive functions distinct from line supervision (responsible for oversight of job execution). Best performed by para-professional management personnel, they are predicated on the principle that maintenance management will achieve best results when each mechanic is given specific tasks to be completed in a definite time period (scheduling) in a specific manner (planning). Mechanics do not plan for their own efficiency.

OBSTACLE **15**
Insufficient Leadership of Job Execution

Maintenance excellence entails proper performance (planning) of the most important jobs (coordination and scheduling), and efficient and reliable execution without wasting resources.

Once jobs have been planned, coordinated, and

scheduled, supervision leads job execution. Supervisors are provided with strategic depictions (the schedule) of all work to be performed by members of their teams (individually and as a unit) throughout the schedule week. Thereby, supervisors, technicians, and internal customers all know the agreed upon work to be completed during the current week. The next job to be performed by each technician is always known and ready.

> *"Maintenance technicians accomplish more when working to achieve published, challenging, and fair expectations."*

Labor, materials, tools, and equipment needs are identified and provided for prior to resources being committed in the field. At that point, supervisors become responsible for the tactical execution of jobs on their schedule with the highest compliance possible, given real-time circumstances that develop throughout the week.

At the time of scheduled execution, the responsible supervisor assigns each job to the appropriate technician as allocated on the schedule. The supervisor then monitors progress throughout the schedule week. Schedules provide the insight necessary to determine when efforts are falling short of expectation and tactical intervention is required.

Technicians and supervisors provide meaningful feedback to the management team (especially planners) to facilitate improvement of future work preparation. Feedback includes compliance and reasons for non-compliance with the weekly schedule. It further identifies areas where planned job packages require refinement.

Related supervisory duties include:

- Check preparedness for each day's work. Despite preparation performed by the planner/scheduler, checking and rechecking assures positive outcome, resultant compliance, and reliability.
- Assign jobs to specific team members based upon both the resource deployment scheme and the order of importance agreed to and depicted on the schedule.
- Control time lost between jobs, at breaks, and during shift changes.
- Spend two-thirds of each day at job sites providing effective leadership throughout execution, follow-up, and feedback.
- Assure that crews reference planned job packages and make full use of preparatory efforts to perform jobs reliably and efficiently. Constant effort is needed to improve work methods, completeness, neatness, quality, and efficiency.
- Remain abreast of schedule status and monitor progress toward schedule milestones.

- Detect when jobs are running into trouble and take tactical actions as appropriate. Early intervention favorably impacts maintenance schedule compliance and operational uptime, quality, and output.
- Sacrifice the least important jobs as necessary to respond to true emergencies (schedule breaks).
- Note description and duration of the schedule breaks.
- When schedule compliance does falter, protect the most important jobs. Endeavor to complete them by end of the week. Then protect as many other jobs on the original schedule as possible (in order of importance).
- Each evening, adjust the schedule as necessary for the balance of the week. Focus on the most important!
- As each job is finished, confirm proper completion in terms of quality, thoroughness, clean up, customer satisfaction, and feedback including a reasonably accurate distribution of time and materials to each job.
- Verify that it was performed according to plan.
- Verify customer satisfaction.
- When a job deviates from plan, determine why.

- Notate the planned job package with appropriate recommendations (additions, corrections, and other improvements).
- Collect planned job packages and return to the planner/scheduler.

- Mark the schedule for calculation of compliance.

- At the end of the schedule week, calculate schedule compliance with the planner/scheduler and review results with team members.

- Schedule compliance and reasons for noncompliance are reviewed during the next coordination meeting.

This supervisory scenario reflects effective management. After the fact excuses for poor execution with low schedule compliance do not.

"The horse is out of the barn."

The Current Supervisory Situation

Supervisors are undervalued. This situation has worsened in popular management philosophies, such as Self-Directed Work Teams. More people report to first line supervision than to any other organizational level. Supervisors are management's direct link to the work force and the focal point for labor productivity and quality workmanship. They provide the leadership that workers need and deserve. Their place is at the job site — leading, motivating, and training their team. Unfortunately, their critical role has been diluted by a variety of other assignments that are more appropriately assigned to planners, engineers, buyers, and administrative assistants.

Maintenance supervisors often lack required leadership and advanced technology skills. Traditionally, they were selected for supervision because they were the best mechanics. Although upward mobility from the work force is positive, selected individuals must have the required aptitude for supervision and must be offered transitional training and coaching.

Advancing equipment technology exacerbates the situation. The skill and equipment knowledge many supervisors brought to the job is not the same as required for today's technology. Just as technicians need related skills training, so do their supervisors. Without it, supervisors are not equipped to guide their teams in today's technical world.

Maintenance supervisors can also be key contributors to the entrenchment of a reactive culture. They and their teams have received most of their praise for prompt response to equipment failures. Often, they have never experienced the benefits of a proactive environment. Maintenance supervisors are as equally set in their reactive ways as production supervisors.

Model the Way; Praise Proactive Accomplishment!

Process Steps to Address These Barriers (continued from previous section)

86. Reinvigorate Supervisory Leadership.

87. Relieve Supervisors of Non-Supervisory Responsibilities.

88. Focus Supervisor Presence and Efforts to the Job Site.

Supervisors should focus on leading and developing their team.

89. Provide Supervisory and Skills Training as Needed.

90. Enforce Adherence to Proper Supervisory Responsibilities.

Follow the guidelines outlined above.

EPILOGUE

In conclusion, I want to emphasize that insights presented within this book must be assimilated by the entire organization, beginning with management at all levels and within all functions. A mindset change is essential.

I am confident that the book will reach the hands of managers within the maintenance function and that they will fully grasp the book's message and purpose. I am less confident, however, that it will reach and be read by upper management or by managers within the operating functions. Accordingly, I encourage each champion of Maintenance and Reliability to make sure that a copy of this book gets into the hands of other key managers. Although short, the book is comprehensive. If managers will invest two hours of reading time, the book will provide a wake-up call as to how their own actions impact the over-arching issues that determine reliability. The return can be enormous.

Finally, I applaud your continuing efforts to reestablish world-class operational competitiveness and leadership. Your efforts are essential to your plant, your company, our economy, and thereby our country. Thank you all! I have enjoyed working with so many of you.

Don Nyman
Hilton Head Island, SC
2009

Appendix

Appendix A
Governing Principles and Concepts

World-Class maintenance functions are structured and built upon very similar concepts, principles, and operational practices that are technically and administratively sound.

To realize Maintenance/Reliability Excellence, the following beliefs must be shared and inculcated throughout the organization:

No.   Principle

1   The excellence process must have sustained management understanding, commitment, support, involvement, and funding; communicated by a well-conceived Maintenance/Reliability Mission Statement.

2   There must be an organization-wide commitment to a proactive maintenance/reliability evolution.

3   A holistic approach to maintenance/reliability excellence must be pursued, rather than pursuing mere maintenance cost reduction.

4   Objectives, Goals, and Targets must be clearly established and disseminated, with effective management reporting and control of progress toward milestone events.

5   The Computerized Maintenance Management Information base must be complete, current, and reliable.

6   The process must be built upon a partnership between the maintenance team, the operating team, and all support teams.

7   All members of the organization must share responsibility for reliability through custodianship of their own equipment and associated work area.

8   Inter-functional responsibilities for reliability and to each other must be clearly established.

9 The maintenance/reliability excellence process must be integrated (21 essential building blocks - see Appendix I).

10 The organization must pursue development of all individuals to capture their full potential and contribution, providing job fulfillment and a sense of accomplishment. A participative environment is essential.

11 All personnel must receive required education and training.

12 Equipment Operators should be trained and certified to perform routine maintenance.

13 Preparation of Planned Work, Leadership of Work Execution, and Engineering dedication to elimination of failures must be formalized.

14 Control Spans for all functions must comply with good practices.

15 Materials Management must effectively support maintenance and reliability efforts.

16 Provision must be made for timely performance of the three types of maintenance work: prompt response to true emergencies, reliable routine service (PM/PdM), and timely relief of backlog.

17 The Reliability Process must be spearheaded by a strong Maintenance/Reliability Engineering function.

18 Operational Plans must provide necessary equipment access as defined by Maintenance/Reliability Plans for each asset.

19 Preventive/Predictive Maintenance must be conducted as a "controlled experiment" with compliance >95%.

20 Workload Backlog must be managed within control limits, regularly balancing resources with workload.

21 Schedule Compliance for backlog relief should be >90%.

Appendix B
The Maintenance/Reliability Iceberg

Principle: A holistic approach to Maintenance/Reliability Excellence must be pursued rather than a narrow pursuit of mere maintenance cost reduction.

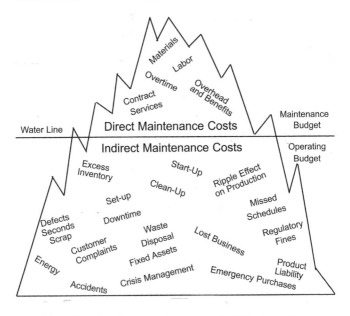

The objective is to optimize profitability, quality, customer service, employee safety, and environmental protection. Simple maintenance cost reduction is often in opposition to this objective.

Appendix C
The Contribution of
Maintenance/Reliability
Excellence

				Appendix C
	Potential Contribution			
				Annual Potential
Budget Item		**Total**	**%**	**Profit Improvement**
		($MM)		($MM)
Maintenance Budget (Above Waterline)				
Labor		$4.0	20%	$0.80
Materials & Parts		$4.0	15%	$0.60
Overtime		$1.0	40%	$0.40
Overhead & Benefits		$1.5	15%	$0.23
Contract Support		$2.0	30%	$0.60
	Subtotal:	**$12.5**		**$2.63**
Operating Budget (Below Waterline)				
Equipment Failures		$6.0	40%	$2.40
Idling & Minor Stoppages		$0.8	40%	$0.32
Off Standard Speed		$0.5	40%	$0.20
Process Defects		$2.0	20%	$0.40
Yield Losses		$3.0	20%	$0.60
Recovery Costs (Remelt etc.)		$0.5	20%	$0.10
Set Ups, Change Overs & Adjustments		$20.0	20%	$4.00
Start Up Costs & Losses		$9.0	20%	$1.80
Inventory Carrying Cost		$0.7	20%	$0.14
Express Delivery Costs		$0.3	40%	$0.12
Demurrage		$0.2	40%	$0.08
Waste Disposal		$0.1	40%	$0.04
Energy Losses		$0.5	40%	$0.20
Impact of Late Deliveries & Other Customer Compliants		$0.6	40%	$0.24
Lost Business (Profit) Opportunity		$3.0	40%	$1.20
Product Liability Costs		$0.7	60%	$0.42
Cost of Environmental Upsets		$0.3	60%	$0.18
Clean Up of Spills		$0.1	40%	$0.04
Costs Associated with Accidents		$0.2	80%	$0.16
	Subtotal:	**$48.5**		**$15.00**
Capital Avoidance		**$20.0**	**40%**	**$8.00**

			Master Plan - Maintenance/Reliability Excellence							Time Line								Appendix D
Chapter	Obstacle	Process Step	Process Activities	Responsibility	J	F	M	A	M	J	J	A	S	O	N	D	Etc.
			PROCESS INITIATION	Management													
		1	Access the Current Culture and Situation														
			o Conduct Assessment Relative to Foundation Beliefs	Facilitator													
			o Conduct Activity Sampling of Maintenance Work Day														
			o Assemble Results and Recommendations														
			CREATING A RELIABILITY CULTURE	Management													
			Initiate the Proactive Reliability Transformation														
			Educate the Key Management Team														
		2	o Present Maintenance/Reliability Excellence Seminar	Facilitator													
	1		o Compare Assessment Results to Foundation Beliefs														
			Confirm the Objective														
			o World-Class Operations Supported by Maintenance/Reliability Excellence														
			Obtain Authorization to continue the Process														
			Establish Maintenance/Reliability Excellence Steering Team	Management													
			STEERING COMMITTEE ACTIONS	Committee													
			Identify the "Champion" of the Maintenance/Reliability Process														
One			Conduct Steering Team Work Shop to Tailor Site-Specific:	Facilitator													
		3	o Provide further Education as needed														
		4 and 5	o Established "Shared Beliefs" to be inculcated throughout the Organization														
		6	o Develop the Overall Vision and Maintenance/Reliability Mission Statements														
		7	o Define Roles and Responsibilities	Committee													
		8	o Develop the Process Master Plan with Responsibilities and Time Lines														
			o Develop Budget necessary to implement plan and build the Partnership														
			Build Justification for Approval														
	2		o Define and Quantify the Maintenance Iceberg														
			o Select Appropriate Maintenance/Reliability Metrics														
			- Identify the sources of Necessary Information														
			- Capture the Historical Baseline														
		10	- Set Ultimate Goals and Progressive Interim Targets														
			- Identify and Define Necessary Management Reporting														
			- Specify System for Reporting Asset Failures/Production Downtime														
		11	Establish Incentives														
		12	Present above Deliverables to Management for Authorization to Continue														
		13	Nurture a Covenantal Partnership among all Functional Organization Units														
		14	Sell Maintenance/Reliability Excellence to All Parties throughout the Organization														
		15	Implement, Coordinate, Monitor and Achieve														
			Begin Trend Reporting	Champion													
	2		o Near Term Interim Targets														
			o Long range Ultimate Goals														
			o Establish Benchmarks														

Continued on next page

			Description	Role
One			o Levels Reflective of Maintenance/Reliability Excellence	
			Measure Results by Comparison to Historic Trends	
			Plot Progress to Master Plan and Dynamically Refine It as Appropriate	Committee
			Initiate Quarterly Steering Team Meetings	Management
			EMPHASIS ON MAINTENANCE/RELIABILITY EXCELLENCE	Planner
	3	16	Purify Current Backlog	Organization
		17	Aggressively Inspect Facility and Equipment	
		18	Bootstrap All Assets Back to Reliability	Champion
		19	Reinvigorate PM/PdM Process as a Controlled Experiment	
		20	Apply Root Cause Analysis to All Failures	Maint/Rel Eng
		21	Applaud, Honor, Reward, Preserve, Share,and Expand Pro-Active Results	
	4		**EFFECT CULTURAL TRANSITION TO PRO-ACTION**	
		22	Drive to 90% Weekly Schedule Compliance	
		23	Realign Expectations	
	5		**STABILIZE THE ORGANIZATION AND SUSTAIN THE MISSION**	Management
		24	Keep Committed Team in Tact	
		25	Limit Organizational Changes	
		26	Assure Viable Succession Plan	
		27	Maintain Focus on Master Plan, Goals, and Targets	
		28	Apply Trend and Correlation Charts	
	6	29	Establish a Senior, Multi-Function Steering Committee	
		30	Develop a Master Plan	
		31	Develop the Necessary Budget to Implement the Plan	Champion
		32	Build the Justification for Approval	
		33	Sell the Plan to All Parties and Obtain Approval	
		34	Conduct All Necessary Education and Training at All Levels	
		35	Implement, Coordinate, Monitor and Achieve	

Continued on next page

Phase	#	Item	Description	Responsible
			STRUCTURE THE ORGANIZATION FOR RELIABILITY	
Two	7	36	Structure Organization to Assure Performance of All Three Work Types	Management
		37	Distribute Resources Approximately 15%/10%/75%	
		38	Establish Planner/Scheduler Position	Champion
		39	Strive for High Schedule Compliance (>90%)	
		40	Provide List of Fill-In Jobs for Response Crew	
		41	Trend Amount of Fill-in Work Accomplished	Planner
		42	Establish Maintenance/Reliability Engineering Position	Management
			RELIABILITY	
Three	8 and 9	43	Project Peak Demand to be Placed on Each Asset	Maint/Rel Eng
		44	Develop Optimal Maintenance Plan for Each Asset	
		45	Quantify Crew Size, Labor-Hours by Skill, and Downtime Duration for Each Asset	Planner
		46	Establish Realistic Capacity of Each Asset	Maint/Rel Eng
		47	Integrate Plans and Needs into Annual Operating, Sales, and Business Plans	Champion
		48	Make Full Use of All Valleys in the Annual Operating Plan	
			BALANCE RESOURCES WITH WORKLOAD	
Four	10	49	Provide for the Demands of All Maintenance Plans	Management
		50	Quantify Workload by Nature	Planner
		51	Regularly (Quarterly) Balance Resources with Workload	Management
			SKILLS TRAINING	
Five	11	52	Define the Skills Required to Preserve Process Reliability (Job Task Analysis)	
		53	Assess Current Skills of Individual Technicians (Training Needs Analysis)	
		54	Document Training Needs to Close the Gap between Required and Existing Skills	Facilitator
		55	o Establish the Training Curriculum, Media, and Sources Thereof	
		56	o Define a Pay-For Skills Progression Model	
		57	Launch Skills Training	
			o Monitor Impact of Skills Training Upon Reliability	Champion
			MATERIALS SUPPORT	
Six	12	58	Establish Strong Buying, Storeroom, and Receiving Functions	Management
		59	Establish a Well-Designed, and Effectively Located Storeroom	
		60	Establish an Inventory Steering Committee	
		61	Integrate Material Control and Work Order Control Modules	IT
		62	Fully Load the Materials Database	
		63	Adhere to Effective Materials/Maintenance Procedures	Materials Mngr/IT
		64	Continuously Refine Material Control Parameters	
		65	Establish Inventory Reservation for Planned Jobs	IT
		66	Establish Material Delivery Procedures	
		67	Establish Effective Vendor Partnership	Materials Mngr

Continued on next page

			PREPARATION FOR EFFECTIVE WORK EXECUTION
Seven	13		**PLANNING**
		68	Establish a Strong Planning Function
			o Conduct Planner/Scheduler Training
		69	Capture All Planning Efforts into Reference Files
		70	Continuously Refine Planned Job Packages
		71	Assure Effective Feedback to planners from Technicians and Supervisors
		72	Refine Work Measurement and Development of Labor Libraries
			o Train All Planners to Estimate in a Consistent Manner
			o Initiate Job Slotting to Support Expansion and Refinement of Estimates
			o Continue to Refine the Estimates Until They Become Worthy Standards
			COORDINATION
		73	Designate Primary Liaisons Within Operations for Coordination With Maintenance
		74	Select Most Logical Schedule Week
		75 and 76	Firmly Establish Timing of Weekly Coordination Meeting and Required Attendees
	14		**SCHEDULING**
			Indoctrinate Maintenance Personnel to New Scheduling Procedures
			Activate Weekly Coordination and Scheduling Meeting
		77	o Emphasize Importance of Conducting PM/PdM as a "Controlled Experiment"
		78	o Emphasize Timely Performance of "Corrective Maintenance"
			Launch the Weekly Scheduling Process
		79	o Fully Load All Technicians to the Weekly Schedule (by job, day and individual)
		80	o Schedule Least Important Jobs to Best Troubleshooters
		81	o Concentrate on the Most Important Jobs
			o Measure and Trend Schedule Compliance
		82	o Identify and Address "Reasons for Non-Compliance"
		83	o Study Recurring Windows of Opportunity
			- If Down ... Do!
		84	o Measure and Trend Crew Efficiency
		85	Regularly Coach Planners in Their Pursuit of Effective Planning and Scheduling
	15		**LEADERSHIP OF JOB EXECUTION**
		86	Reinvigorate Supervisory Leadership
		87	Relieve Them of Non-Supervisory Responsibilities
		88	Focus Their Presence and Efforts to the Job Site
		89	Provide Supervisory Skills Training As Needed
		90	Engrain Adherence to Supervisory Responsibilities

	Appendix E
Seminar Outline	
Maintenance/Reliability Excellence	
Subject/Discussion Point	**Reference/Comment**
Introduction	
Explanation of Seminar Title	
Supportive of World-Class Operation	
Integration - of the Essential Building Blocks	Appendix I - Arch
Management	
Partnership	
The Ills of Reactive & Deferred Maintenance	
Where the Maintenance Dollar Goes	
Convincing Management	
Top Ten Reasons	
Process Foundations	
Ingredients for Cultural Transition	Obstacle 4 & 5
Vision & Mission Statements	Appendices E & G - Obstacle 2
Governing Principals & Concepts - Shared Beliefs	Appendix A - Obstacle 1
Partnership	
Custodianship	
Inter-Functional Responsibilities (Each to the Others)	
Operator Responsibilities within Team Concepts	
Commitment - Organization Wide	
Holistic Excellence vs. Simplistic Cost Reduction	Appendices B & C - Iceberg - Obstacle 3
Assessing the Current Situation	
What is Your Current State?	
Survey of Maintenance Issues	
Evaluation	
Array of Results	
Improvement Potential	
Building the Organizational Team	
The Organizational Environment	
The Three Key Functions & Associated Spans of Control	
Balancing Resources with Workload	Appendix J - Work Program - Obstacle 10
Three Types of Maintenance Workload	
Required Staffing	
As Dictated by Backlog	
Build Up by Workload	
Build Up by Budget Item	
Distribution by Work Type	
Organize for Pro-Action not Reaction	Appendix F - Org Chart - Obstacle 7
- Work Type Structure	
- Other Workable Structures (Qualified)	
- Those not Recommended	
Associated Budget	Obstacle 6

Continued on next page

Skills Training	
The Skills Crisis	
Education/Training Options	
Pay-for-Skills	
Progression Paths	Obstacle 11
Succession Planning	
The Essential Informational Base	
Work Order Coding	
Asset Configuration	
Work Management Process (WMP)	
Materials Support	
Objectives	Obstacle 12
Statistical Inventory Control (SIC)	
Just-In-Time (JIT)	
Stock Classification	
ABC Analysis	
Reliability Centered Maintenance (RCM)	
Preventive, Predictive & Programmed Routines	
Fixed, Floating & Metered Schedules	
Balancing the Schedule	
Conducting It as a "Controlled Experiment"	
Inspection & Correction	
Patterns of Equipment Failure	
Determination of Realistic Asset Capacity	Obstacle 8
Engineering Required Maintenance Plan for Each Asset	Obstacle 9
Work Preparation	
Planning	Obstacle 13
Criteria of a Planned Job	
Why Bother to Plan?	
Who is in the best position to effectively Plan	
The Planning Process	
Work Measurement	
Forms of Work Measurement	
Analytical Estimating	
Benchmarks for Slotting New Jobs	
Coordination	
Agreement as to Importantance	
Establishing Expectation for the Schedule Week	Obstacle 14
Scheduling	
Schedule Format	Appendix K
Presentation of the Schedule	
Schedule Compliance	
Reasons for Non-Compliance	
Leadership of Work Execution	Obstacle 15
How Maintenance Supervisors Spend The Day - Currently	
Recognizing Importance of Maintenance Supervisors	
Job Leadership and Team Development	

Appendix F 1 Organizational Table Reference Functions

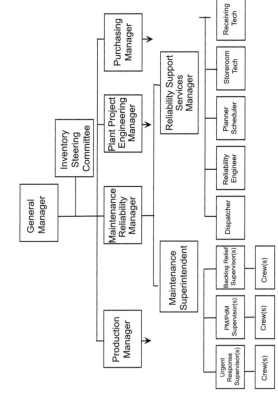

Appendix F 2 Organizational Structure by Maintenance Work Type

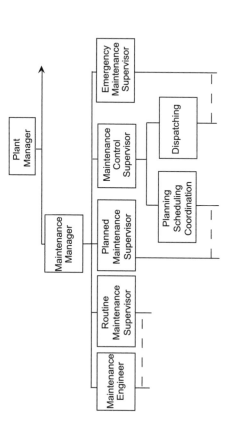

Appendix G
Vision Statement

Our vision is to be a world-class producer of all product lines that we offer to our customer population. Our quality, customer service, cost, safety, and environmental protection will be equal to or better than any competitor in the world offering a comparable product. Reliability will be emphasized in all endeavors!

Appendix H
The Maintenance/Reliability Mission Statement

To provide timely, quality, and cost effective service and technical guidance in support of short-range and long-range operating plans. Insuring through pro-action, rather than reaction, that assets are maintained to support required levels of reliability, availability, output capacity, quality, and customer service. This vision is to be fulfilled within a working environment that fosters safety, high morale, and job fulfillment for all members of the operational team while protecting the surrounding environment and well-being of consumers.

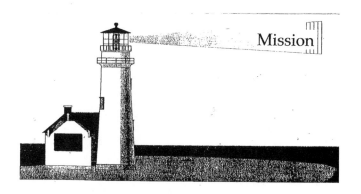

Mission

Appendix I
The Maintenance Excellence Arch
The Road to Reliability

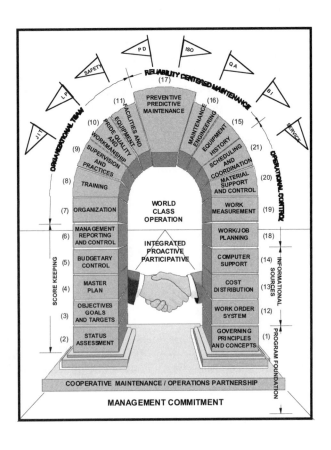

Explanation of the Maintenance Arch

The arch reflects integration of twenty-one build-ing blocks essential to achievement of reliability excel-lence. It is built upon bedrock of "sustained management commitment, support, involvement, and funding. Upon the bedrock is formed a footer reflecting the essential "Operations/Maintenance Partnership". Neither function can achieve reliability without support from each other as well as from the entire organization.

The banners flying from the Arch reflect the many management initiatives that are maintenance dependent (Just-in-Time, Loss Prevention, Employee Safety, Avoidance of Property Damage, Certification, Quality Assurance, Avoidance of Business Interruption, and Customer Service).

The building blocks are interdependent. It is of lit-tle sustainable benefit to undertake improvement of one without addressing needs of the others:

- Successful endeavors begin with shared beliefs (Governing Principles). Status Assessment establish-es the current state relative to beliefs. The Master Plan sets forth actions, resources, responsibilities, and time lines required to close the gap between current state and established Objectives, Goals, and Targets. Budget refinement is required to support the Master Plan.

- Preventive/Predictive Maintenance is the keystone of the Arch and is the vehicle by which reliability is assured. It requires a strong Maintenance/Reliability Engineering function using Equipment History and deploying Root Cause Failure Analysis to optimize PM/PdM routines.
- Organization must support proactive reliability not simply reactive response. Skills Training and Facilities, Tools and Equipment in which technicians can take Pride are essential if Supervision is to achieve functional Quality Assurance and adherence to established policies and procedures.
- Planning and Materials Support are essential for Maintenance is to properly perform the right work in the first instance. Work Measurement, Coordination, and Scheduling establish the expectations for each week.
- The Computerized Work Order System with reliable Cost Distribution feed into Management Reporting, thereby providing the feedback by which sustained commitment is earned.

Reliability requires total commitment, partnership, integration, and pro-activity resulting in cultural change. Independent changes do not lead to reformation of the business entity. It is not just the independent building blocks, but also the manner in which they are erected. A number of integrated and harmonious changes are required to achieve a quantum leap forward.

The Maintenance Work Program
Balancing Resources with Workload

Available Resources

Crew size: 20

Straight Time Labor Hours per Week	800
Planned Overtime per Week	96
Labor Hours Contracted or Borrowed Per Week	0
Total Labor Hours Available per Week	896
Less Indirect Commitments (Weekly Averages)	
Lunch (if paid)	0
Vacation	120
Absence	24
Training	56
Meetings	40
Special Assignments	40
Average Labor Hours Loaned to other areas	40
Other Indirect	10
Total Indirect Labor Hours Projected per Week	330
Total Labor Hours per Week Available for Direct Work	566
Commitments Other Than Backlog Relief (Weekly Averages)	
Emergency/Urgent (unschedulable)	100
Routine PM	120
Other Fixed Routine Assignments	0
Sub Total	220
Net Resources Available for Backlog Relief	346

Backlog Data	Current		Backlog Weeks	Current	Target
Backlog Labor Hours in Ready Status	3200		Ready	9.2	2 to 4
Total Labor Hours of Backlog	4800		Total	13.9	4 to 6

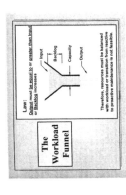

The Workload Funnel

Law:
Output must be equal to or greater than input or Backlog increases

Input
Backlog
Capacity
Output

Therefore, resources must be balanced with workload or transition from reactive to proactive maintenance is not feasible.

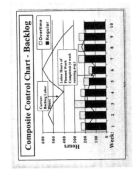

Composite Control Chart - Backlog

☐ Overtime   ■ Regular

Hours
600
500
400
300
200
100
0

Current Backlog Labor Hours

Labor Hours of Planned Work Completed (4 week running avg.)

Week: 1 2 3 4 5 6 7 8 9 10

Appendix K

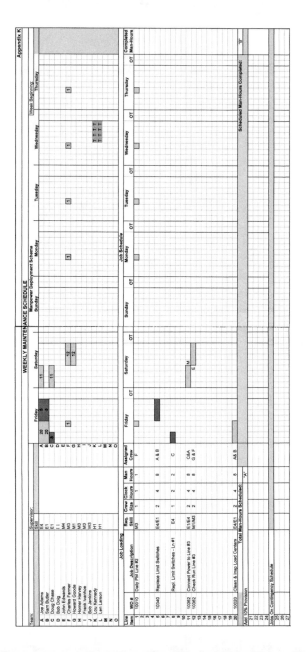

Appendix L

Appendix L   Sources of Support

Source	Form Of Support	City	Contact	Phone	Web Site
MPACT Learning Center	B D	Greensboro, NC	Pete Little	336-379-1444	MPACTLearning.com
People and Processes	A	Yulee, FL	Dave Bertolini	843-814-3798	peopleandprocesses.com
MET DEMAND	A B C	Charleston, IL	Kevin Lewton	888-427-4330	metdemand.com
Norman and Associates	A B C	Charleston, SC	Andy Norman	843-442-7667	NormanandAssociates.com
Life Cycle Engineering	A B C	Charleston, SC	Tom Dabbs	843-744-7110	LCE.com

Forms of Support

A Assessment
B Maintenance/Reliability Excellence Seminar
C Master Plan Facilitation
D Skills Training

Appendix M
Managerial Obstacles to Reliability

1. Lack of Understanding, Belief, and Commitment Shared by the Entire Organization
2. Lack of Integrated Missions, Plans, and Incentives
3. Emphasis on Cost Reduction Rather Than Reliability Excellence
4. Failure to Effect Cultural Transition from Reaction to Pro-Action with Realignment of Expectations
5. Insufficient Organizational Stability to Sustain the Reliability Mission
6. Lack of a Master Plan, Associated Budget, and Commitment
7. Maintenance Functions Organized for Reactive Response
8. No Clear Quantification of Asset Capacity Required to Reliably Support Business Plans
9. No Clear Quantification of Maintenance Workload or Downtime Required to Preserve Capacity
10. Imbalance of Maintenance Resources with Maintenance Workload
11. Skills of Many Maintenance Personnel Fall Short of Levels Required to Maintain Advancing Equipment Technology
12. Procurement is Focused Too much on Initial Cost vs. Life Cycle Cost
13. Insufficient Preparation for the Execution of Maintenance Work
14. Insufficient Statement of Expectations
15. Insufficient Leadership of Job Execution

Index